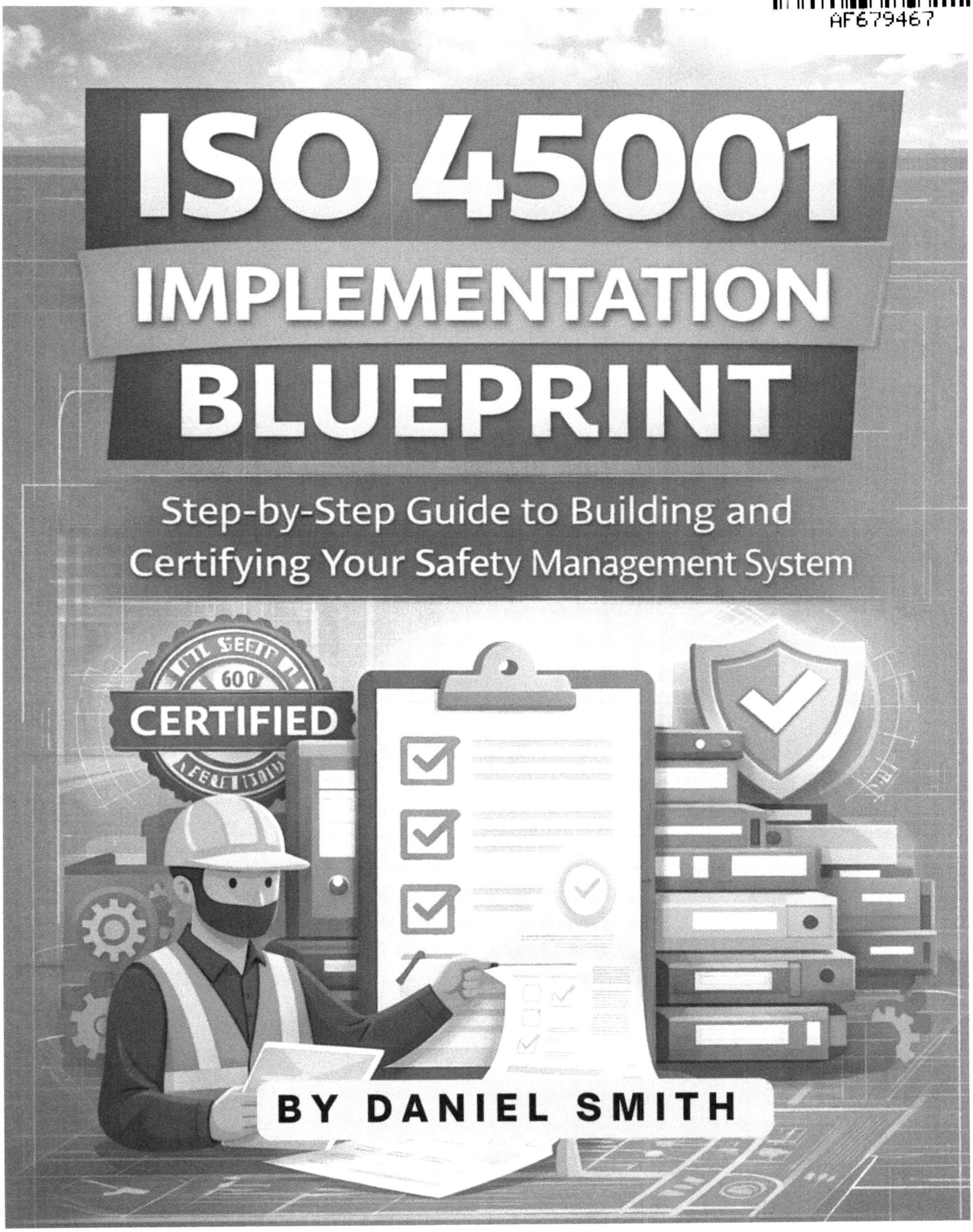
ISO 45001
IMPLEMENTATION
BLUEPRINT
Step-by-Step Guide to Building and
Certifying Your Safety Management System
CERTIFIED
BY DANIEL SMITH

ISO 45001 Implementation Blueprint

Step-by-Step Guide to Building and Certifying Your Safety Management System

By Daniel Smith

Copyright Page

Dedication Page

To all the businesses and professionals working to strengthen information security, protect data, and build trust. Your dedication to confidentiality, integrity, and availability doesn't just meet ISO 27001 — it raises the standard for everyone.

About the Author

Daniel Smith is a recognized authority on international standards and management systems. With more than two decades of experience, he has supported organizations of all sizes in successfully achieving and maintaining certification.

Broad Expertise

His work spans a wide range of ISO standards, including quality management, information security, occupational health and safety, environmental management, and many others. This broad expertise allows him to offer guidance that is relevant across industries and business functions.

Practical and Clear Guidance

Through his books, toolkits, and resources, Daniel simplifies complex requirements into clear, step-by-step methods. His practical approach helps professionals move from theory to real-world application, ensuring compliance and measurable improvement.

Trusted Worldwide

Daniel's resources are trusted by managers, consultants, and business leaders around the world. His mission is to provide practical knowledge and tools that empower organizations to reduce risks, strengthen systems, and achieve sustainable success through certification.

Offer Information

Want to get the full document?

We are offering the Word document version of this book at $20 for those who want to take the content with them wherever they go. For more information and to receive your copy, please contact us at:

Email: standardbookshub@gmail.com

Table of Contents

Chapter 1

Gap Analysis and Implementation Planning

Overview of ISO 45001

ISO 45001 is an international standard that outlines the requirements for an Occupational Health and Safety Management System (OHSMS). Its purpose is to provide organizations with a systematic approach to managing safety and health risks in the workplace, ensuring the safety of workers and other stakeholders. It aims to reduce accidents, injuries, and illnesses, while promoting a culture of safety within organizations of all sizes and across various sectors.

ISO 45001 was developed by the International Organization for Standardization (ISO) and was published in March 2018. It is the first global standard focusing on occupational health and safety (OHS) that is designed to integrate with other management system standards, such as ISO 9001 (Quality Management) and ISO 14001 (Environmental Management).

The standard is structured around the **High-Level Structure (HLS)**, which is common to many ISO management standards. This ensures that it can be easily integrated with other management systems that an organization may already be using. The framework of ISO 45001 provides guidelines on risk management, safety policies, and continual improvement processes. It helps organizations create a safe work environment by identifying hazards, assessing risks, implementing control measures, and continuously improving their safety practices.

Key Elements of ISO 45001

ISO 45001 is structured into **10 clauses**, with each providing essential guidance for a comprehensive OHSMS. Here's a breakdown of the key elements:

1. **Clause 1: Scope**
   Defines the scope of the standard, specifying the requirements for establishing, implementing, maintaining, and continually improving an OHSMS within the context of the organization's overall business.
2. **Clause 2: Normative References**
   This clause refers to any other documents that are essential for applying ISO 45001 effectively. It primarily points to ISO 9000, the fundamental principles and vocabulary standard.
3. **Clause 3: Terms and Definitions**
   Provides the necessary definitions for terms used throughout the standard to ensure clarity and consistency in understanding.

4. **Clause 4: Context of the Organization**
   Emphasizes the importance of understanding the internal and external factors that affect the organization, including legal and regulatory requirements, stakeholder needs, and the broader health and safety environment. This clause also includes the identification of risks and opportunities that need to be addressed.
5. **Clause 5: Leadership and Worker Participation**
   This clause stresses the role of top management in fostering a safety culture, demonstrating commitment, and ensuring that there is active worker participation in the OHSMS. It includes requirements for assigning roles and responsibilities for safety-related tasks.
6. **Clause 6: Planning**
   The planning section covers hazard identification, risk assessment, and the creation of objectives and actions to eliminate or minimize risks. It requires organizations to establish a safety policy, set goals for safety improvements, and develop plans to meet those goals.
7. **Clause 7: Support**
   This clause focuses on the resources, training, communication, and documentation required to establish an effective OHSMS. It also discusses the need for documented information to support the system and provide clarity for all stakeholders.
8. **Clause 8: Operation**
   This section requires organizations to implement the processes and controls necessary to address the identified hazards and manage risks. It includes detailed guidelines for operational procedures, emergency preparedness, and response measures.
9. **Clause 9: Performance Evaluation**
   The evaluation process involves monitoring, measurement, analysis, and assessment of OHSMS performance. It includes regular internal audits and reviews to ensure compliance and continual improvement. Key performance indicators (KPIs) and performance evaluations help assess the effectiveness of safety measures.
10. **Clause 10: Improvement**
    This final clause outlines the process for identifying nonconformities and taking corrective actions. It emphasizes the need for continuous improvement in health and safety practices to adapt to changing conditions and to prevent future incidents.

Why ISO 45001 is Important

The implementation of ISO 45001 has numerous benefits for organizations, employees, and other stakeholders. These benefits include:

1. **Improved Workplace Safety**
   ISO 45001 helps reduce workplace incidents, injuries, and illnesses by identifying and addressing potential risks proactively. This leads to safer working conditions and a reduction in the number of accidents.
2. **Legal Compliance**
   Organizations that comply with ISO 45001 are more likely to meet legal requirements related to occupational health and safety. This helps avoid legal penalties, fines, and lawsuits due to non-compliance.
3. **Reduced Costs**
   By preventing accidents and injuries, ISO 45001 can help reduce the costs associated with workplace incidents, such as medical expenses, insurance premiums, and lost productivity. Additionally, it can lower compensation claims and other liability costs.
4. **Enhanced Employee Well-being and Engagement**
   Implementing ISO 45001 demonstrates an organization's commitment to the health and safety of its employees. This can lead to improved employee morale, engagement, and retention.
5. **Improved Reputation and Brand Image**
   Organizations that are certified under ISO 45001 can enhance their reputation by showing a commitment to the health, safety, and well-being of their employees. This can improve relationships with customers, partners, and other stakeholders.
6. **Continuous Improvement and Risk Management**
   ISO 45001 promotes a culture of continuous improvement and ensures that safety measures evolve with changing risks. The standard encourages organizations to evaluate and mitigate risks on an ongoing basis, ensuring a proactive approach to safety.
7. **Global Standard for Health and Safety**
   ISO 45001 is an internationally recognized standard, which means that organizations certified under this system can demonstrate their commitment to global safety practices. It enhances an organization's competitive advantage in international markets.

Key Benefits of Implementing ISO 45001

- **Legal Compliance**: Helps organizations comply with occupational health and safety laws and regulations, avoiding fines and legal actions.
- **Improved Worker Morale**: By focusing on employee safety, organizations create a safer work environment, leading to higher job satisfaction and reduced turnover.
- **Risk Management**: Provides a structured approach to managing workplace hazards and mitigating risks, which leads to fewer accidents and injuries.
- **Cost Savings**: Reduces the financial impact of workplace incidents, including insurance premiums and medical costs, by preventing injuries and illnesses.
- **Reputation and Trust**: An ISO 45001 certification enhances the organization's credibility and fosters trust with clients, partners, and employees.

Conclusion of Introduction

ISO 45001 is a comprehensive framework that helps organizations identify and mitigate risks to create a safer and healthier workplace. Through a systematic approach, it ensures that an organization's safety management practices align with global standards, providing a solid foundation for continuous improvement and enhanced safety performance. Understanding the full scope of ISO 45001 is the first step toward implementing it effectively, and it forms the basis for the gap analysis and implementation planning in subsequent chapters.

Why Gap Analysis is Critical to Success

Gap analysis is a fundamental step in the implementation of any management system, including **ISO 45001**. It helps organizations identify the difference between their current occupational health and safety (OHS) management practices and the requirements of the ISO 45001 standard. Conducting a thorough gap analysis enables organizations to determine the areas where they are already compliant with the standard and where they need to make improvements to meet the requirements.

A **gap analysis** provides several critical benefits that make it essential to the success of ISO 45001 implementation. One of the key reasons for its importance is that it offers a clear, structured path for the organization to align its OHS practices with the standard's requirements. By identifying areas of non-compliance or gaps in current procedures, policies, and practices, an organization can take targeted actions to bridge these gaps before seeking ISO 45001 certification.

This process starts with reviewing the existing occupational health and safety management system (OHSMS) against the ISO 45001 clauses. The gaps identified could range from minor procedural issues to larger systemic weaknesses. The analysis is not limited to a one-time assessment but should be an ongoing process that helps track the effectiveness of implemented changes and provides continuous improvements.

The gap analysis process typically involves the following key steps:

1. **Review of Current Systems and Practices**
   Organizations must assess their current health and safety management systems to understand where they stand. This includes evaluating existing policies, procedures, risk assessments, and compliance with local legal regulations. It is important to conduct this review holistically to get a clear picture of the organization's current state of safety practices.
2. **Identification of Gaps**
   Once the current system has been reviewed, the next step is to identify any gaps in alignment with ISO 45001 requirements. These gaps could be in various areas, such as leadership commitment, employee participation, risk management processes, operational controls, emergency preparedness, or monitoring and auditing systems. Identifying these gaps allows the organization to understand the areas that need attention and improvement.
3. **Prioritization of Gaps**
   Not all gaps are equal, and some may have more significant implications for safety and compliance than others. The gap analysis process helps prioritize which areas should be addressed first. Typically, high-priority gaps relate to areas where there is a direct risk to worker safety or where compliance with legal regulations is not being met.
4. **Developing Action Plans for Improvement**
   Once the gaps have been identified and prioritized, organizations can develop actionable plans to address them. These plans include timelines, resources, responsibilities, and any necessary training or updates to policies and procedures. The action plan should be specific, measurable, achievable,

relevant, and time-bound (SMART), ensuring that improvements are tracked and the gap is successfully closed.

5. **Ensuring Continuous Monitoring and Improvement**
   A gap analysis is not a one-off process. It should be repeated periodically to ensure that the OHSMS remains aligned with ISO 45001 and continues to meet safety objectives. The process helps keep the system dynamic, addressing new risks and emerging challenges as the workplace evolves. Continuous monitoring and improvement ensure that safety practices evolve with organizational growth and changes in the regulatory environment.

The gap analysis process also provides the foundation for setting specific **goals** and **objectives** in the **ISO 45001 implementation plan**. It helps determine the key performance indicators (KPIs) to monitor and evaluate the success of the implemented changes. Without this initial analysis, organizations could face difficulties in prioritizing resources, determining timelines, and avoiding inefficiencies in meeting ISO 45001 requirements.

A successful gap analysis ensures that the OHSMS is built on a solid foundation, and it helps to maintain ongoing improvements. By addressing the gaps identified, organizations enhance their safety culture, mitigate risks, and provide a safer working environment for employees. This step sets the stage for the next phase of ISO 45001 implementation—**developing policies, procedures, and SOPs**, which will be aligned with the insights from the gap analysis.

In summary, gap analysis is critical for identifying the gaps between an organization's current safety practices and the ISO 45001 standard. It enables organizations to focus on areas that need improvement, prioritize safety efforts, and ensure that the implementation of ISO 45001 is successful, efficient, and impactful in creating a safer workplace.

The Importance of a Well-Structured Implementation Plan

A well-structured implementation plan is essential for the successful deployment and certification of **ISO 45001** within an organization. This plan serves as the roadmap, guiding the organization through the entire process of aligning its occupational health and safety management system (OHSMS) with the requirements of the standard. It ensures that all necessary steps are taken in an organized manner, deadlines are met, resources are allocated effectively, and progress is continuously monitored.

Here's why a **well-structured implementation plan** is crucial:

1. **Clear Direction and Focus**
   A structured plan helps establish clear goals, objectives, and milestones, ensuring that all efforts are aligned with ISO 45001 requirements. It provides direction to the team and keeps everyone focused on achieving specific outcomes, reducing the chances of inefficiency and confusion.
2. **Resource Allocation and Budgeting**
   Implementing ISO 45001 requires both human and financial resources. A detailed plan helps identify the resources needed for each phase, ensuring that the right expertise, tools, and equipment are available. Budgeting is also crucial, as it helps organizations plan for costs related to training, system upgrades, audit preparations, and certification fees.
3. **Timeline Management**
   An implementation plan includes a timeline with realistic deadlines for each step of the process. This ensures that the implementation stays on track and meets the desired timeline for ISO 45001 certification. It also helps avoid delays and prevents last-minute rushes, ensuring a smooth, organized process.
4. **Roles and Responsibilities**
   A key element of a structured plan is the assignment of clear roles and responsibilities. It outlines who is accountable for each task, which promotes accountability and transparency. Assigning responsibility also ensures that there is a dedicated person or team working on each aspect of the implementation process.
5. **Risk Management and Mitigation**
   The implementation plan helps identify potential risks early in the process. By addressing these risks proactively, organizations can develop mitigation strategies to avoid disruptions and delays in achieving certification. A well-structured plan includes contingency measures in case of unforeseen challenges.
6. **Progress Monitoring and Evaluation**
   The implementation plan allows organizations to track their progress against the defined objectives. Regular progress evaluations help identify issues

early on and ensure that the project stays within scope. Continuous monitoring also ensures that any changes or adjustments can be made to improve the process and ensure successful completion.

7. **Alignment with Organizational Goals**
   A well-organized plan aligns the implementation of ISO 45001 with broader organizational goals. It integrates safety objectives with the company's overall mission, ensuring that safety is prioritized as part of the organization's long-term strategy for growth and sustainability.
8. **Employee Engagement and Participation**
   Involving employees at every stage of the planning and implementation process is key to success. A structured plan ensures that the workforce is engaged, trained, and motivated throughout the process. This also fosters a culture of safety, where everyone is committed to creating a safer working environment.

Table: The Importance of a Well-Structured Implementation Plan

Aspect	Explanation	Impact
Clear Direction and Focus	Establishes clear goals, objectives, and milestones aligned with ISO 45001 requirements.	Ensures efforts are focused and organized, reducing inefficiencies and confusion during implementation.
Resource Allocation and Budgeting	Identifies necessary resources (human, financial, and technical) for successful ISO 45001 deployment.	Ensures sufficient resources are available for each phase, preventing resource shortages and controlling costs.
Timeline Management	Includes deadlines for each phase of implementation, ensuring a structured and timely approach.	Keeps the process on track, ensuring that deadlines are met and preventing delays.
Roles and Responsibilities	Assigns clear responsibilities for each task or phase in the process.	Promotes accountability and transparency, ensuring tasks are completed effectively by the right people.

Risk Management and Mitigation	Identifies potential risks and defines mitigation strategies for managing them.	Helps prevent or manage disruptions, reducing the likelihood of major setbacks and ensuring the process stays on track.
Progress Monitoring and Evaluation	Allows for continuous tracking of progress against objectives and milestones.	Provides insights into the implementation's success, enabling necessary adjustments and corrective actions to ensure timely certification.
Alignment with Organizational Goals	Ensures the implementation is in line with broader organizational objectives and mission.	Aligns safety initiatives with the organization's long-term strategy, reinforcing the importance of safety at all levels.
Employee Engagement and Participation	Promotes employee involvement and training to ensure engagement in the safety culture.	Fosters a collaborative approach to safety, ensuring active participation from all levels of the workforce and enhancing the impact of safety measures.

In conclusion, a well-structured **ISO 45001 implementation plan** provides a clear framework for success. It ensures that all necessary steps are taken in the right order, with the appropriate resources, timelines, and personnel in place. This leads to a smoother, more efficient implementation process, reduces risks, and helps ensure the organization meets ISO 45001 certification requirements in a timely and organized manner.

Key Concepts

Gap Analysis Overview

A **gap analysis** is a diagnostic tool that helps organizations evaluate their existing Occupational Health and Safety Management System (OHSMS) in relation to the requirements of the ISO 45001 standard. The purpose of the gap analysis is to identify the differences or "gaps" between what the organization is currently doing in terms of health and safety management and what is required to comply with the standard. This analysis forms the basis for planning the necessary changes to ensure the OHSMS meets ISO 45001 certification criteria.

The gap analysis process provides a comprehensive understanding of where the organization's safety management system stands and what needs to be improved to align with ISO 45001. It also offers the organization a roadmap for making improvements and preparing for the certification audit. The ultimate goal is to close any gaps and ensure that all aspects of the OHSMS are compliant with the ISO 45001 standard.

Identifying Gaps in Safety Management Systems

Identifying gaps in an organization's current safety management system is the core of the gap analysis. These gaps can be related to various aspects of the existing safety practices, procedures, and overall system structure. The most common gaps include:

1. **Non-compliance with Legal or Regulatory Requirements**
   The organization may not be adhering to specific health and safety laws or regulations that are part of the ISO 45001 standard.
2. **Lack of Documentation or Formalized Procedures**
   The absence of documented policies, procedures, or safety protocols can be a significant gap, as ISO 45001 requires thorough documentation and clear safety processes.
3. **Ineffective Risk Assessment Processes**
   The organization may not have a formal or systematic approach to identifying and assessing workplace hazards and risks, which is a critical requirement of ISO 45001.
4. **Insufficient Employee Engagement and Participation**
   A key element of ISO 45001 is involving workers in the OHSMS. If the organization has not effectively engaged employees in safety initiatives, this represents a gap that must be addressed.
5. **Poor Emergency Preparedness**
   If the organization lacks adequate emergency response plans or does not

regularly conduct drills, it presents a safety gap that needs immediate attention.

Key Areas for Evaluation: Compliance, Risk Management, and Safety Culture

To conduct a thorough gap analysis, there are several key areas that need to be evaluated:

1. **Compliance**
   Assessing compliance is the first step in identifying gaps. This involves reviewing the organization's adherence to legal and regulatory safety requirements, such as labor laws, local safety regulations, and industry standards. ISO 45001 requires organizations to go beyond mere legal compliance and to implement best practices in health and safety.
2. **Risk Management**
   Risk management is a core component of ISO 45001. The organization must evaluate its processes for identifying, assessing, and mitigating workplace hazards. A comprehensive risk management system should include risk assessments, control measures, and regular reviews to ensure that safety risks are effectively managed.
3. **Safety Culture**
   An organization's safety culture determines how seriously safety is taken by employees at all levels. It is essential to assess the organization's safety culture to ensure that safety is integrated into daily activities and is supported by management. ISO 45001 requires a proactive safety culture where safety is prioritized and employees are engaged in safety decisions.

How to Conduct a Gap Analysis

Conducting a gap analysis involves several systematic steps to evaluate the organization's existing safety practices against ISO 45001 requirements. The process helps uncover areas that need improvement and provides a clear picture of how to proceed with the implementation of the OHSMS.

Steps for Assessing Existing Systems

The following steps outline how to assess the existing OHSMS during a gap analysis:

1. **Review the Current OHSMS Documentation**
   Begin by gathering and reviewing all the relevant safety policies, procedures, SOPs, and risk assessments in place. This documentation will serve as the foundation for evaluating the current system.
2. **Conduct Interviews and Surveys**
   Engage key stakeholders, including management, safety officers, and employees, to understand how safety practices are implemented at the operational level. Interviews and surveys will provide insight into the practical application of safety policies.
3. **Conduct Workplace Inspections and Audits**
   Perform workplace inspections to assess the effectiveness of safety measures in real-time. This includes reviewing machinery, equipment, and work processes for potential hazards.
4. **Evaluate Safety Performance Metrics**
   Analyze existing safety performance indicators, such as accident rates, near misses, and worker feedback, to gauge the effectiveness of current safety management practices.

Mapping Current Practices to ISO 45001 Requirements

Once the organization's current practices are assessed, the next step is to map these practices against the ISO 45001 standard. This is a critical step in identifying where the gaps are. The mapping process involves reviewing each clause of ISO 45001 and comparing it with the organization's existing OHSMS practices. For example:

- **ISO 45001 Clause 4 (Context of the Organization)** requires understanding the internal and external factors that affect the OHSMS. The organization must assess whether its current safety management system considers these factors adequately.
- **ISO 45001 Clause 5 (Leadership and Worker Participation)** emphasizes the role of management and employee participation in the OHSMS. The

organization should evaluate if workers are actively engaged in safety discussions and if management is demonstrating a commitment to safety.

By mapping current practices to the ISO 45001 clauses, the organization can identify areas where changes or improvements are necessary to align with the standard.

Risk and Opportunity Assessment

Identifying Risks in Workplace Safety

Identifying risks in the workplace is a critical part of the gap analysis. This involves reviewing potential hazards and assessing their impact on employee safety. Some common workplace safety risks include:

- **Physical Hazards**: Machinery, equipment, and physical work environments that pose a risk of injury.
- **Chemical Hazards**: Exposure to toxic chemicals or substances.
- **Ergonomic Hazards**: Workstations or tasks that lead to repetitive strain injuries.
- **Psychosocial Hazards**: Stress, harassment, and mental health concerns that can affect employee well-being.

By identifying these risks early on, the organization can take proactive steps to mitigate them and align with ISO 45001's risk management requirements.

Evaluating Opportunities for Improvement

In addition to identifying risks, the gap analysis process should also focus on evaluating **opportunities for improvement**. These opportunities could include:

- **Streamlining Safety Processes**: Reviewing safety procedures to remove inefficiencies and make them more effective.
- **Improving Employee Engagement**: Increasing employee involvement in safety decisions and providing better communication channels for safety concerns.

- **Leveraging Technology**: Implementing safety management software or tools to track incidents, inspections, and compliance more efficiently.

By identifying and acting on these opportunities, the organization can continuously improve its safety practices, in line with ISO 45001's emphasis on continuous improvement.

Implementation Planning

Once the gaps have been identified and opportunities for improvement have been evaluated, the next step is **implementation planning**. A well-organized implementation plan ensures that the organization takes actionable steps to close the gaps and align its OHSMS with ISO 45001.

Setting Clear Objectives and Deliverables

The implementation plan must outline clear, measurable objectives. These objectives should directly address the identified gaps and focus on improving specific areas of the OHSMS. Each objective must have defined deliverables, which can be tracked to ensure progress.

Developing a Timeline for ISO 45001 Implementation

Creating a timeline helps the organization stay on track and meet key milestones. A well-structured timeline outlines the steps and deadlines for each phase of the implementation, from gap analysis to certification.

Resource Allocation and Budgeting

Effective resource allocation is essential for the success of ISO 45001 implementation. The plan should outline the human, financial, and technical resources needed to achieve the objectives. A budget should be developed to cover costs such as training, system upgrades, and audit preparations.

Assigning Roles and Responsibilities

To ensure accountability and transparency, the implementation plan should clearly define roles and responsibilities. Assigning specific tasks to individuals or teams ensures that each aspect of the implementation is addressed and managed appropriately.

Practical Tools

1. **Gap Analysis Template**
   A template that can be used to document and evaluate current safety management practices against ISO 45001 requirements. It includes sections for identifying gaps, prioritizing them, and developing action plans for improvement.
2. **Risk and Opportunity Register**
   A tool for tracking identified risks and opportunities in the workplace. It helps the organization evaluate and document safety hazards and areas for improvement, ensuring that they are addressed in the implementation plan.
3. **Action Plan Template for ISO 45001**
   A template for creating a detailed action plan that outlines the steps needed to close the identified gaps and achieve ISO 45001 compliance. It includes timelines, resources, and responsibilities for each action.

In conclusion, conducting a thorough **gap analysis** and **implementation planning** process is critical to successfully achieving ISO 45001 certification. By identifying gaps in safety management practices and implementing a structured plan with clear objectives and deadlines, organizations can effectively address weaknesses, engage employees, and ensure compliance with the standard.

Chapter 2

Developing Policies, Procedures, and SOPs

Introduction

In the journey towards achieving ISO 45001 certification, one of the most crucial steps is the development of policies, procedures, and Standard Operating Procedures (SOPs). These elements form the foundation of the organization's Occupational Health and Safety Management System (OHSMS). They are the primary tools used to ensure that the organization is meeting both the regulatory and ISO 45001 standards, ensuring a safe and healthy working environment.

This chapter will delve into the role of these documents in ISO 45001 implementation, how they contribute to compliance, and the continuous improvement of workplace safety.

The Role of Policies, Procedures, and SOPs in ISO 45001

1. **Occupational Health and Safety Policy (OHS Policy)**

The **Occupational Health and Safety Policy** is the cornerstone of an organization's commitment to worker health and safety. It establishes the organization's overall health and safety goals, the framework for achieving them, and a commitment to continual improvement. The policy should be clear, concise, and signed by top management to ensure that leadership's commitment to health and safety is demonstrated.

In ISO 45001, the policy should:

- **Reflect the organization's commitment** to providing a safe working environment and preventing accidents, injuries, and ill health.
- Be aligned with the organization's overall business strategy.

- Set measurable objectives for OHSMS implementation, which are reviewed periodically to ensure relevance.
- Ensure compliance with all legal and regulatory safety requirements.
- Promote the continuous improvement of safety standards.

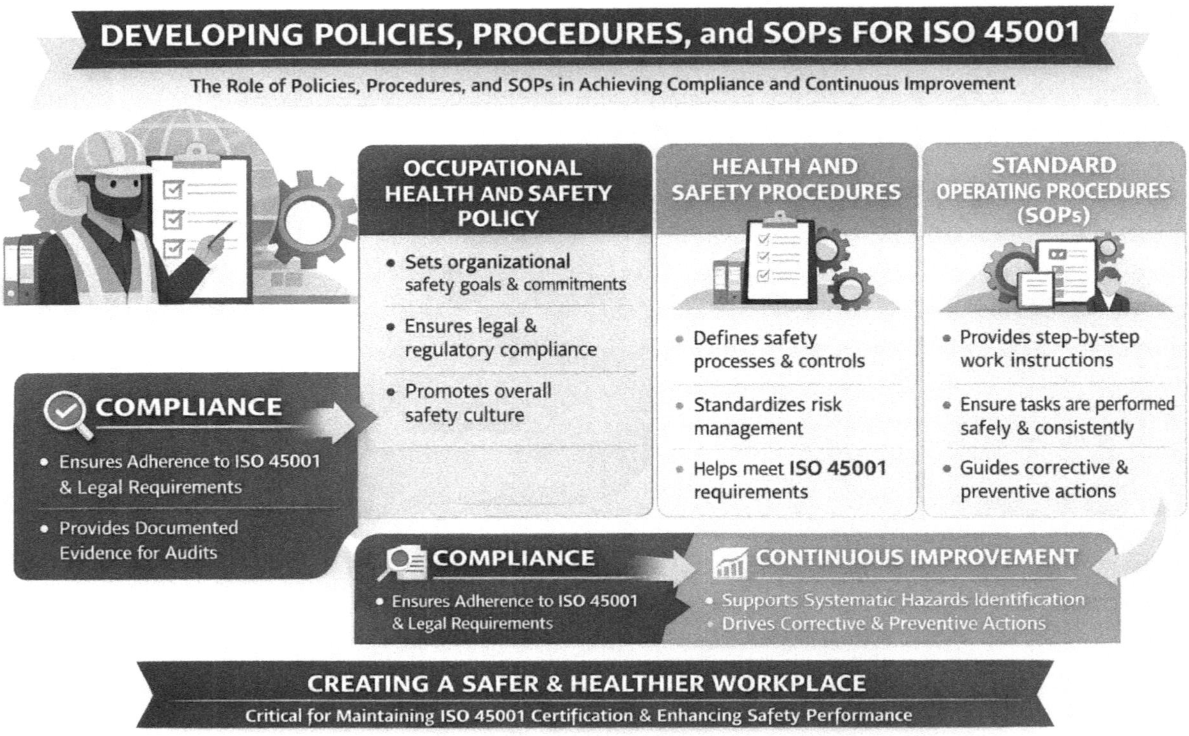

The OHS Policy sets the tone for the entire OHSMS by emphasizing top management's commitment to the welfare of employees. It serves as a guide for the development of more detailed procedures and SOPs.

2. **Health and Safety Procedures**

Procedures in ISO 45001 outline specific steps and actions required to meet the OHS policy's objectives. These are the "what" and "how" of safety practices. Each procedure describes the operational actions to be followed to manage particular safety risks, such as fire safety, hazardous material handling, or equipment maintenance.

Procedures help in:

- **Standardizing tasks**: Ensuring that every worker follows the same process, which increases consistency and reduces errors.
- **Documenting operational controls**: Establishing clear instructions for risk management and control measures.
- **Ensuring compliance**: Procedures help the organization meet legal obligations, industry standards, and ISO 45001's requirements.
- **Identifying key risks**: Procedures are specifically designed to manage safety risks associated with specific tasks or work areas.

For example, a procedure for **handling hazardous chemicals** would describe the steps involved in safely storing, using, and disposing of chemicals, the safety equipment required, and the steps to follow in the event of a spill.

3. **Standard Operating Procedures (SOPs)**

SOPs are detailed, step-by-step instructions designed to ensure that tasks are carried out correctly and consistently. SOPs are important for tasks that are performed frequently and that could pose significant risks if done incorrectly.

SOPs should:

- **Provide detailed instructions**: Each step of the process is broken down, ensuring that the procedure is carried out consistently and safely every time.
- **Include safety considerations**: SOPs address potential hazards associated with the task and detail the safety measures needed to mitigate these risks.
- **Ensure operational efficiency**: By standardizing tasks, SOPs help increase productivity and reduce the likelihood of errors or accidents.
- **Be regularly reviewed**: SOPs must be updated periodically to reflect changes in legislation, equipment, or processes.

For instance, a **lockout/tagout** SOP for equipment maintenance ensures that machines are properly shut off and cannot be restarted while workers are repairing or maintaining them. This procedure reduces the risk of injury due to unexpected machine startup.

How They Contribute to Compliance and Continuous Improvement

1. **Contribution to Compliance**

ISO 45001 requires that organizations have a comprehensive approach to managing occupational health and safety risks, which includes complying with local and international legal and regulatory requirements. Developing policies, procedures, and SOPs plays a significant role in ensuring that the organization meets these requirements.

- **Regulatory Compliance**: The organization's policies and procedures help ensure that all health and safety risks are managed according to legal standards. For example, the procedures for handling hazardous substances will reflect the relevant health and safety laws concerning the storage, handling, and disposal of chemicals.
- **Internal Consistency**: With detailed policies and procedures in place, the organization ensures that everyone in the company follows the same safety protocols, reducing confusion and variation in safety practices.
- **Auditing and Monitoring**: Well-documented procedures and SOPs make internal audits easier to conduct. They provide auditors with concrete evidence of how the organization manages safety, facilitating compliance audits and certifications.

2. **Contribution to Continuous Improvement**

ISO 45001 emphasizes **continuous improvement** as a key component of a successful OHSMS. Policies, procedures, and SOPs serve as tools for monitoring and improving health and safety practices over time.

- **Systematic Approach to Risk Management**: Procedures and SOPs help to identify hazards, assess risks, and implement control measures. The continuous monitoring of these processes enables organizations to identify areas where safety could be improved and to take corrective actions.
- **Feedback and Updates**: As new safety issues emerge or as legal regulations change, policies, procedures, and SOPs must be updated. Continuous feedback from employees, audits, and accident investigations can lead to improvements in safety procedures. For example, if a particular procedure is

identified as ineffective or causing delays, the SOP can be revised to improve its efficiency and effectiveness.

- **Employee Engagement and Participation**: Involving employees in the creation and review of policies and procedures ensures that safety practices are practical and relevant. Employee input helps improve the safety culture and ensures that the policies and procedures are supported by those who are directly affected by them.
- **Monitoring Effectiveness**: Regularly reviewing the implementation of policies, procedures, and SOPs helps track the effectiveness of the safety system. If certain procedures are found to be ineffective in preventing injuries or accidents, they can be adjusted to address those shortcomings.
- **Performance Evaluation**: ISO 45001 requires organizations to measure and evaluate the effectiveness of their OHSMS. Well-documented procedures provide data that can be used to assess performance and identify areas for improvement. KPIs (Key Performance Indicators) related to safety can be monitored, and new procedures can be implemented when necessary to meet safety goals.
- **Corrective and Preventive Actions (CAPA)**: When incidents or near-misses occur, the organization must investigate the root causes. Policies, procedures, and SOPs ensure that corrective actions are taken to address identified issues. Additionally, preventive actions can be implemented to avoid similar incidents in the future.

Conclusion

In ISO 45001, policies, procedures, and SOPs are not just documentation requirements but essential tools for establishing, maintaining, and improving the health and safety management system. They ensure that organizations not only comply with legal and regulatory safety requirements but also establish a culture of safety that supports continuous improvement.

By developing clear, effective policies, procedures, and SOPs, organizations can prevent workplace accidents, protect employees' health, reduce operational risks, and ultimately improve their OHS performance. These documents lay the groundwork for a strong OHSMS and serve as the foundation for achieving ISO 45001 certification and fostering a safe working environment.

Key Concepts

Developing an Occupational Health and Safety Policy

An **Occupational Health and Safety Policy (OHS Policy)** is a critical document in any organization that sets the foundation for its health and safety management system. This policy serves as a formal declaration of the organization's commitment to creating a safe and healthy working environment, and it guides the development of subsequent safety procedures and practices. The OHS Policy must align with the organization's overall goals and culture while ensuring compliance with legal and regulatory requirements.

The development of an OHS policy typically follows these steps:

- **Management Commitment**: The policy must be signed and endorsed by top management to demonstrate leadership's commitment to worker safety and health.
- **Clear and Concise Language**: It should clearly outline the organization's safety objectives, goals, and safety-related commitments.
- **Stakeholder Involvement**: It should reflect the needs and concerns of all stakeholders, including workers, regulatory bodies, and customers.
- **Legal Compliance**: The policy must ensure adherence to applicable local, regional, and international health and safety laws and regulations.
- **Continuous Improvement**: It should emphasize the importance of continuous safety improvements and risk management.

A strong OHS Policy demonstrates leadership's commitment to a safety culture, clearly communicates safety goals, and sets a framework for maintaining a safe work environment.

Elements of a Strong Safety Policy

A **strong safety policy** should contain several key elements:

1. **Safety Vision and Commitment**: Top management's dedication to safety should be stated upfront, emphasizing that safety is a priority at all organizational levels.
2. **Clear Safety Goals**: The policy should define specific safety objectives, such as reducing accident rates, improving training programs, or enhancing safety culture.
3. **Roles and Responsibilities**: The policy must outline the roles and responsibilities of both management and employees in maintaining safety standards.
4. **Compliance with Laws and Regulations**: It should ensure the organization complies with all local, national, and international safety standards and regulations.
5. **Worker Involvement**: The policy should promote active employee participation in safety decision-making and encourage reporting of hazards or unsafe conditions.
6. **Continuous Improvement**: The policy should promote a commitment to improving safety processes, conducting regular risk assessments, and implementing corrective actions as necessary.

A well-designed policy should not only be a legal document but also a tool for fostering a proactive safety culture.

Ensuring Management Commitment

Management commitment is a crucial element in the success of any Occupational Health and Safety Management System (OHSMS). Top management must lead by example and demonstrate a genuine commitment to maintaining a safe work environment. This can be achieved through:

- **Visible Leadership**: Management should be actively involved in safety initiatives, regularly visiting work sites, and encouraging employee participation in safety programs.

- **Resource Allocation**: Allocating sufficient resources for safety-related activities, including personnel, equipment, and training.
- **Setting the Tone**: Management should set the expectations for safety performance and ensure that safety is incorporated into all decision-making processes.
- **Encouraging Open Communication**: Promoting an environment where employees feel comfortable discussing safety concerns without fear of retaliation.

When management is committed, it positively impacts the entire organization's approach to health and safety, setting a foundation for a strong safety culture.

Creating Standard Operating Procedures (SOPs)

Standard Operating Procedures (SOPs) are detailed, step-by-step instructions that describe how specific tasks or processes should be carried out. SOPs are essential for ensuring that tasks are completed consistently, efficiently, and safely. They provide clarity to employees about how to safely perform their duties, ensuring that risk factors are minimized and legal and procedural requirements are met.

Key aspects of creating effective SOPs include:

- **Detailed Instructions**: SOPs should clearly define the steps for performing a task and ensure that safety risks are identified and mitigated.
- **User-Friendly Format**: SOPs must be easy to read, with straightforward language and a structured format that can be followed without confusion.
- **Relevant Safety Protocols**: Safety measures must be incorporated into the SOP to minimize hazards and ensure safe execution.
- **Training**: Employees must be trained to understand and follow SOPs to maintain high safety standards.
- **Review and Update**: SOPs should be reviewed and updated regularly to reflect changes in tasks, equipment, or safety regulations.

Structuring Effective SOPs

An effective SOP should follow a clear structure to ensure it is both comprehensive and accessible. The key components of an SOP include:

1. **Title**: Clearly identifies the procedure and its scope.
2. **Purpose**: Explains the reason for the SOP and the specific task it covers.
3. **Scope**: Defines the boundaries of the SOP, including where it applies and who is responsible for following it.
4. **Responsibilities**: Lists the roles and individuals who are responsible for carrying out the procedure.
5. **Step-by-Step Instructions**: Provides detailed steps for completing the task, ensuring clarity and ease of understanding.
6. **Safety Considerations**: Identifies any potential hazards and outlines the safety measures to mitigate these risks.
7. **Documentation and Records**: Specifies the documentation requirements and records that need to be maintained.
8. **Review and Revision**: Details the process for reviewing and revising the SOP, ensuring it stays up to date.

Ensuring Clear Communication of Safety Expectations

Clear communication of safety expectations is essential to the success of an OHSMS. The organization should establish:

- **Safety Meetings**: Regular safety meetings to communicate safety updates, share lessons learned, and discuss potential hazards.
- **Training Programs**: Structured training sessions to educate employees on safety policies, procedures, and emergency protocols.
- **Posters and Signage**: Visible safety posters, warning signs, and hazard indicators throughout the workplace to remind employees of key safety measures.
- **Open Channels for Feedback**: Ensuring employees can easily report hazards or suggest improvements to the safety system.

Effective communication is a key component of a strong safety culture, fostering engagement, transparency, and a proactive approach to safety.

Procedures for Risk Assessment, Incident Reporting, and Corrective Actions

1. **Risk Assessment Methodologies**

Risk assessment is a systematic process used to identify hazards, evaluate the risks associated with them, and implement control measures to mitigate those risks. Common risk assessment methodologies include:

- **Qualitative Risk Assessment**: Involves subjective analysis of risk factors based on their likelihood and potential consequences.
- **Quantitative Risk Assessment**: Involves the use of numerical data to assess the probability and impact of risks.
- **Bowtie Analysis**: A method used to identify hazards and visualize control measures, focusing on both preventive and mitigation strategies.
- **HAZOP (Hazard and Operability Study)**: A structured technique used to evaluate risks in complex systems and identify design flaws or operational hazards.

2. **Reporting and Investigation Protocols**

Incident reporting is essential for identifying and addressing safety hazards. Employees should be encouraged to report any accidents, near misses, or unsafe conditions immediately. A structured protocol for incident investigation includes:

- **Incident Reporting Forms**: Easy-to-use forms for reporting incidents, accidents, or unsafe practices.
- **Investigation Team**: A team of qualified individuals who investigate the root causes of incidents and determine corrective actions.
- **Documentation**: Detailed documentation of the investigation process, including root cause analysis, corrective actions, and preventive measures.

3. **Corrective and Preventive Action (CAPA)**

Corrective and Preventive Action (CAPA) is a structured approach to identifying the root cause of incidents, addressing the problem, and preventing future occurrences. The CAPA process includes:

- **Root Cause Analysis**: Identifying the underlying causes of incidents, rather than just addressing the symptoms.
- **Corrective Action**: Taking steps to correct the issue and prevent its recurrence, such as changes in processes, training, or equipment.
- **Preventive Action**: Implementing measures to prevent potential risks from materializing, such as modifying existing procedures or introducing new safety controls.

Table: Key Aspects of Risk Assessment, Incident Reporting, and CAPA

Component	Explanation	Action/Outcome
Risk Assessment Methodologies	Identifying hazards, evaluating risk, and implementing controls.	Ensures safety risks are recognized and managed, preventing accidents and injuries.
Reporting Protocol	A structured system for reporting incidents, accidents, and unsafe conditions.	Promotes early identification of safety issues, enabling prompt corrective actions.
Incident Investigation	A process for identifying the root causes of incidents and implementing corrective measures.	Helps prevent similar incidents in the future by addressing underlying causes and improving processes.
Corrective Action	Immediate steps taken to address the root cause of an incident or safety issue.	Reduces recurrence of incidents by addressing the issues at their source.
Preventive Action	Measures taken to eliminate potential risks before they result in incidents.	Prevents safety incidents by identifying and addressing potential hazards early.

In conclusion, the development of strong **Occupational Health and Safety Policies, procedures**, and **SOPs** is fundamental for ensuring compliance with ISO 45001 and promoting a culture of continuous improvement in workplace safety. These documents guide the organization in managing health and safety risks, providing clear safety expectations to employees, and establishing a system for proactive risk management, incident reporting, and corrective actions.

Practical Tools for ISO 45001 Implementation

Here are three essential templates that can be used to streamline the implementation of your Occupational Health and Safety Management System (OHSMS). These templates are designed to help create a **Safety Policy**, develop **Standard Operating Procedures (SOPs)** for key safety processes, and conduct **Risk Assessments**. These ready-to-use templates will assist in aligning your safety practices with ISO 45001 requirements.

1. Safety Policy Template

This **Safety Policy Template** provides a framework for developing your organization's Occupational Health and Safety Policy. The policy should be tailored to reflect your organization's commitment to safety, compliance with legal regulations, and the promotion of a safety culture.

[Company Name] Safety Policy

1. Policy Statement
At **[Company Name]**, we are committed to ensuring the health, safety, and well-being of all our employees, contractors, and visitors. We will provide and maintain a safe working environment by adhering to the highest standards of occupational health and safety, as outlined in ISO 45001.

2. Safety Objectives
Our safety objectives include:

- Reducing workplace injuries and illnesses.
- Promoting a culture of safety through education, training, and engagement.

- Ensuring compliance with all applicable health and safety laws, regulations, and ISO 45001 standards.
- Continuously improving our safety management practices through audits, reviews, and corrective actions.

3. Scope of the Policy
This policy applies to all employees, contractors, and visitors across **[Company Name]**. It covers all operational areas, including office spaces, production facilities, and fieldwork sites.

4. Management Commitment
Our management is fully committed to the implementation and maintenance of this safety policy. We will allocate the necessary resources to ensure that our OHSMS is effective and continually improved.

5. Employee Participation
We believe that the safety and health of our employees is a shared responsibility. We encourage active participation from all employees in safety meetings, risk assessments, and the reporting of unsafe practices.

6. Review and Revision
This policy will be reviewed annually to ensure its continued relevance and effectiveness. Any changes in legislation or operational requirements will be incorporated into the policy during the review process.

Signed,
[Name of CEO/Manager]
[Position]
[Date]

---

2. SOP Template for Key Safety Processes

The **Standard Operating Procedure (SOP) Template** provides a structured approach to documenting and communicating how specific safety tasks should be performed within your organization. Below is an example SOP for **Fire Safety Procedures**.

[Company Name] Fire Safety SOP

1. Purpose
This SOP outlines the procedures to be followed in the event of a fire emergency to ensure the safety of all employees, contractors, and visitors within **[Company Name]** premises.

2. Scope
This SOP applies to all employees and visitors within **[Company Name]**'s premises, including offices, warehouses, and manufacturing sites.

3. Responsibilities

- **Fire Safety Officer**: Ensure the implementation and monitoring of fire safety measures.
- **Employees**: Familiarize themselves with fire exits, alarms, and emergency procedures.
- **Visitors**: Follow the fire safety instructions as communicated by the host.

4. Fire Safety Procedure

- **Step 1**: If a fire is detected, activate the nearest fire alarm.
- **Step 2**: Evacuate the building using the designated fire exits.
- **Step 3**: Assemble at the designated assembly point and conduct a headcount.
- **Step 4**: Do not use elevators during evacuation.
- **Step 5**: If safe to do so, use fire extinguishers to control small fires.
- **Step 6**: Wait for emergency services and follow their instructions.

5. Emergency Contacts

- Fire Department: [Phone Number]
- Emergency Evacuation Coordinator: [Name & Phone Number]
- Fire Safety Officer: [Name & Phone Number]

6. Equipment and Resources

- Fire extinguishers located at [locations].
- Fire exit signs and routes clearly marked.
- Emergency lighting installed in corridors and stairwells.

7. Review
This SOP will be reviewed annually or after a fire drill or actual incident to ensure its effectiveness.

---

3. Risk Assessment Form

This **Risk Assessment Form** template helps in identifying, evaluating, and controlling workplace hazards. It ensures compliance with ISO 45001's requirement for risk management and helps in systematically addressing safety risks.

[Company Name] Risk Assessment Form

1. Risk Assessment Title
[Provide a brief title/description of the risk being assessed.]

2. Assessment Date
[Insert Date of Assessment]

3. Department/Area
[Specify the department or work area being assessed.]

4. Assessor(s)
[List the names and roles of the persons conducting the assessment.]

5. Hazard Identification

Hazard	Potential Risk	Persons at Risk	Likelihood (High/Medium/Low)	Severity (High/Medium/Low)	Risk Rating (Likelihood x Severity)
[Insert Hazard]	[Insert Potential Risk]	[Workers, Visitors, etc.]	[High/Medium/Low]	[High/Medium/Low]	[Calculated Rating]

6. Risk Control Measures

Control Measure	Person Responsible	Completion Date
[Insert control measure to reduce risk]	[Name/Role]	[Date]

7. Additional Comments/Recommendations

[Provide any additional comments or recommendations for improving safety.]

8. Review and Follow-up

- **Review Date**: [Insert date for review of the effectiveness of control measures.]
- **Follow-up Actions**: [Insert follow-up actions or checks to ensure the risk is adequately controlled.]

Assessor Signature: [Signature]
Manager Approval: [Signature]

Conclusion

These templates provide the foundational documents needed for implementing a compliant and efficient **OHSMS** as part of ISO 45001 certification. The **Safety Policy** outlines the organization's commitment to safety, **SOPs** provide detailed procedures for safety processes, and the **Risk Assessment Form** helps identify and mitigate risks systematically. Using these templates ensures the documentation required for ISO 45001 is complete, actionable, and aligned with safety best practices.

Chapter 3

Operational Controls and Emergency Preparedness

Introduction

In ISO 45001, operational controls and emergency preparedness are fundamental components of the **Occupational Health and Safety Management System (OHSMS)**. These elements ensure that organizations can effectively manage health and safety risks in daily operations and respond promptly to emergencies, minimizing harm to workers and ensuring compliance with safety standards.

This chapter explores the importance of implementing **operational controls**—which are proactive measures aimed at minimizing risks—and **emergency preparedness**, which outlines how to handle potential emergencies when they arise. Together, these elements are key to providing a safe work environment, reducing incidents, and ensuring that the organization is equipped to handle unexpected safety challenges.

The Importance of Operational Controls in ISO 45001

Operational controls refer to the policies, procedures, and practices put in place to manage and mitigate health and safety risks in the workplace. They are proactive measures that help ensure that workplace hazards are identified and effectively controlled before they can cause harm.

Operational controls are a cornerstone of ISO 45001 for several reasons:

1. **Risk Mitigation**
   Operational controls aim to reduce risks by implementing preventive measures to minimize or eliminate hazards before they lead to injuries or incidents. These controls might include engineering solutions (e.g.,

machinery safeguards), administrative controls (e.g., work procedures), or personal protective equipment (PPE).

2. **Compliance with Regulatory Standards**
One of the key objectives of operational controls is to ensure compliance with both legal and ISO 45001 requirements. ISO 45001 mandates that organizations identify hazards and put in place systems that can prevent accidents, which often means meeting or exceeding local health and safety regulations. Operational controls are vital in achieving and maintaining this compliance.
3. **Creating a Safety Culture**
By establishing clear and effective operational controls, the organization demonstrates its commitment to safety at all levels. This approach helps build a culture where safety is prioritized, leading to better overall awareness and worker engagement in safety practices.
4. **Consistency in Safety Practices**
Operational controls ensure that all tasks are performed consistently, following the same safety procedures. This reduces the potential for errors or unsafe practices, which can occur when safety protocols are not followed or are inconsistently applied. By standardizing safety practices, organizations can minimize variation in how safety is managed.
5. **Improvement in Efficiency and Productivity**
Although operational controls are implemented to ensure safety, they also improve operational efficiency. By identifying hazards early, organizations can address them before they lead to accidents or downtime. Effective controls help workers perform tasks safely and efficiently, reducing the potential for disruptions to operations.

Examples of Operational Controls include:

- **Engineering Controls**: Safeguards, machine guards, ventilation systems.
- **Administrative Controls**: Work schedules, job rotation, training programs.
- **PPE (Personal Protective Equipment)**: Gloves, helmets, respirators.
- **Workplace Design**: Layout improvements to minimize hazards, such as ergonomic desk setups.

Together, these controls ensure that risks are actively managed in the workplace, making them a critical part of a successful **OHSMS**.

How Emergency Preparedness Contributes to Worker Safety

Emergency preparedness refers to the planning and actions taken in advance to respond to potential emergency situations in the workplace. These situations can include fires, chemical spills, natural disasters, or workplace violence. Effective emergency preparedness plans not only help organizations respond promptly to emergencies but also protect workers' health and safety, ensuring they are able to act quickly and decisively when needed.

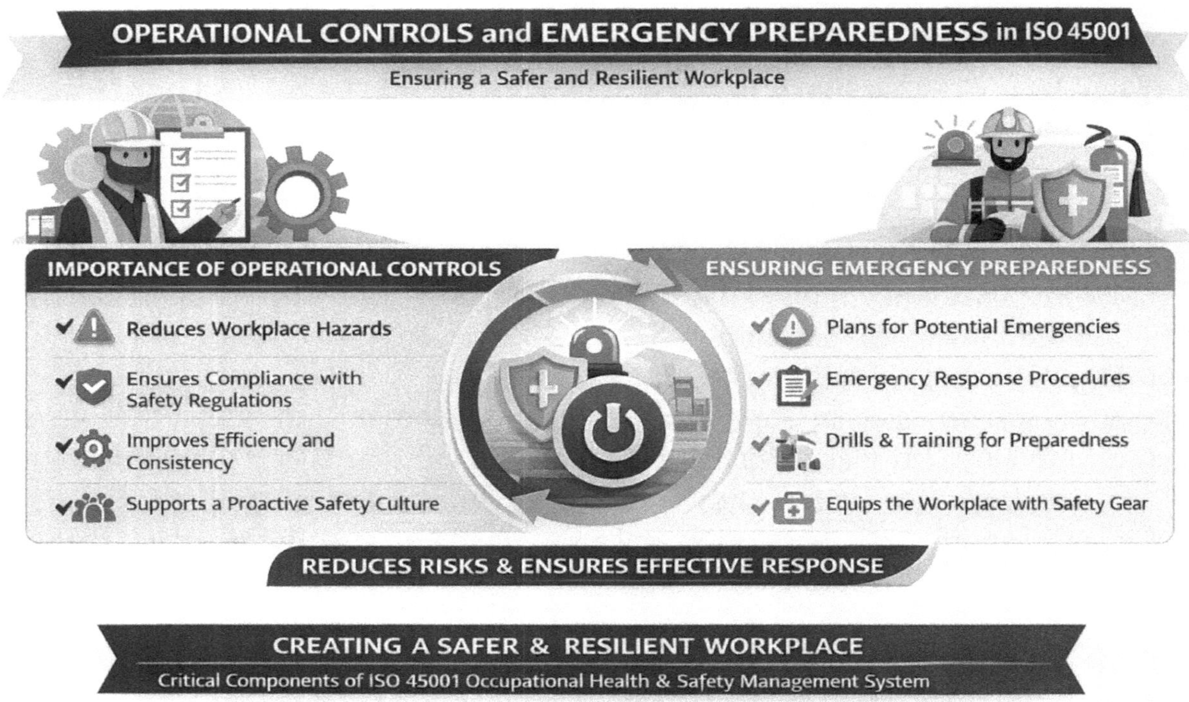

The importance of emergency preparedness in ISO 45001 cannot be overstated. Here's how it contributes to worker safety:

1. **Minimizing Harm During Emergencies**
   A well-developed emergency preparedness plan ensures that all workers know what to do in the event of an emergency, which can significantly

reduce the potential for injuries or fatalities. It provides clear instructions, procedures, and roles to ensure that workers and emergency responders can act swiftly and efficiently.

2. **Training and Familiarity**
   Regular emergency preparedness training ensures that workers are familiar with emergency procedures, such as evacuation routes, first aid procedures, and emergency contact information. This familiarity helps workers stay calm and take appropriate actions during an emergency, reducing the risk of panic or confusion that can exacerbate the situation.
3. **Faster Response Time**
   When an emergency occurs, the faster the response, the lower the risk to worker safety. Emergency preparedness ensures that all employees are trained to respond effectively, and it provides the necessary resources and support, such as emergency kits and first aid equipment. This enables a quick and coordinated response to emergencies.
4. **Compliance with Legal Requirements**
   Emergency preparedness is often a legal requirement in many industries. OSHA and other safety regulatory bodies require that organizations develop emergency response plans, conduct drills, and ensure that the workplace is equipped to handle emergencies. ISO 45001 includes these legal requirements within its framework, ensuring that organizations comply with both local laws and international standards.
5. **Reduced Downtime and Property Damage**
   Having an emergency preparedness plan in place can help reduce operational downtime by ensuring that workers are evacuated safely, equipment is secured, and damage is minimized. In the case of a fire, for instance, a quick evacuation can save lives and prevent extensive damage to equipment or property, enabling the organization to recover quickly after the incident.
6. **Preventing Future Incidents**
   Emergency preparedness is not just about responding to a crisis but also about learning from it. After an emergency, organizations should conduct an analysis of how the situation was handled, identify any weaknesses in the response, and improve the preparedness plan accordingly. This continuous feedback loop contributes to a culture of continuous improvement in safety management.

7. **Protection of Critical Operations**
   An effective emergency plan ensures that business-critical operations can be quickly restored after an emergency. This can involve everything from implementing business continuity plans to setting up backup systems and processes. By safeguarding critical functions, emergency preparedness helps maintain worker safety and organizational stability.

Key Elements of an Effective Emergency Preparedness Plan

A successful emergency preparedness plan consists of several key elements, all of which are designed to ensure that the organization is ready to respond to any emergency that may arise:

1. **Risk Assessment**
   Identifying and analyzing potential emergencies that could affect the workplace. This includes considering factors like fire hazards, chemical spills, natural disasters, and workplace violence.
2. **Emergency Response Procedures**
   Clear procedures outlining what employees should do in the event of an emergency. These should include evacuation routes, assembly points, and detailed instructions for specific emergencies.
3. **Roles and Responsibilities**
   Defining the roles and responsibilities of individuals during an emergency, such as fire wardens, first aid responders, and emergency coordinators. Each person should know their tasks and duties in advance.
4. **Emergency Equipment**
   Ensuring that the workplace is equipped with the necessary emergency response equipment, such as fire extinguishers, first aid kits, emergency alarms, and communication tools.
5. **Training and Drills**
   Regular emergency drills and training sessions to ensure that workers are familiar with emergency procedures and are prepared to respond swiftly in a real emergency situation.
6. **Communication Protocols**
   Clear communication channels should be established, both internally (within the organization) and externally (with emergency responders). Workers

should know how to communicate during an emergency, and key personnel should have the means to quickly reach others.

7. **Post-Emergency Recovery**
   Procedures for assessing and recovering from the emergency, including medical attention for injured workers, a review of the incident, and the restoration of operations.

Conclusion

Both **operational controls** and **emergency preparedness** are critical components of ISO 45001 and contribute significantly to worker safety. **Operational controls** are proactive measures aimed at identifying and managing safety risks before they lead to accidents, while **emergency preparedness** ensures that the organization is ready to respond to emergencies effectively when they occur. By integrating these two elements into the OHSMS, organizations can create a safer, more resilient work environment and ensure compliance with ISO 45001 standards.

Key Concepts

Identifying Operational Controls

Operational controls refer to the specific procedures, safeguards, and mechanisms put in place to minimize or eliminate risks associated with workplace hazards. These controls help reduce the likelihood of accidents, injuries, and illnesses in the workplace. Identifying operational controls involves assessing risks and determining the most effective methods to mitigate these risks.

The process begins with:

- **Identifying hazards** in the workplace (e.g., machinery, chemicals, ergonomics).
- **Evaluating risks** associated with these hazards (i.e., the likelihood and severity of harm).
- **Implementing controls** that reduce these risks to an acceptable level.

Effective operational controls may involve engineering solutions, administrative measures, or the use of personal protective equipment (PPE).

Hierarchy of Control Measures (Elimination, Substitution, Engineering Controls, etc.)

The **Hierarchy of Control Measures** is a widely accepted framework used to determine the most effective control measures for managing health and safety risks. It categorizes control measures from the most effective to the least effective, guiding organizations in their efforts to protect workers.

1. **Elimination**
   The most effective control is eliminating the hazard entirely. This could involve removing dangerous machinery, discontinuing harmful chemicals, or reengineering processes that pose risks. If the hazard can be eliminated, this is the first choice.
   - Example: Replacing a toxic substance with a safer one or removing a dangerous task altogether.
2. **Substitution**
   Substitution involves replacing a hazardous material or process with a safer alternative. This is less effective than elimination but still reduces the risk significantly.
   - Example: Replacing a harmful chemical solvent with a less toxic one or using water-based paints instead of solvent-based ones.
3. **Engineering Controls**
   Engineering controls involve modifying the physical work environment or equipment to reduce exposure to hazards. These controls do not eliminate the hazard, but they can significantly reduce the risk of injury.
   - Example: Installing machine guards, ventilation systems, or noise-dampening devices to control exposure to mechanical hazards, airborne particles, or noise.
4. **Administrative Controls**
   Administrative controls involve altering work practices to reduce exposure to hazards. This might include changing work procedures, establishing safety rules, or creating rotation schedules to limit the amount of time workers are exposed to a specific hazard.

   - Example: Implementing work shifts or breaks to reduce repetitive strain injuries or establishing work procedures that minimize exposure to hazardous chemicals.

5. **Personal Protective Equipment (PPE)**
   PPE is the least effective form of control and should be used as a last line of defense when other controls cannot fully eliminate or control the hazard. PPE includes items like helmets, gloves, goggles, hearing protection, and respirators.
   - Example: Providing workers with gloves, safety goggles, and masks when handling hazardous chemicals or working with machinery.

Implementing Control Measures in the Workplace

Once operational controls are identified, the next step is to implement them in the workplace. This process involves integrating the identified controls into daily operations to ensure that risks are minimized effectively. Implementation can include:

- **Installing physical safeguards**: For example, installing machine guards or automatic shut-off systems.
- **Creating written procedures**: Standard Operating Procedures (SOPs) that clearly describe how to handle hazardous tasks safely.
- **Training employees**: Providing proper training to workers on how to use new controls, PPE, and follow safety procedures.
- **Regularly reviewing controls**: Conducting regular audits and assessments to ensure that controls are effective and being followed.

Emergency Preparedness and Response Plans

Emergency preparedness is critical to ensuring that an organization can respond effectively to unexpected incidents, minimizing harm to workers and preventing further damage. A well-developed **emergency response plan** outlines the procedures for responding to emergencies such as fires, chemical spills, natural disasters, or workplace violence.

Elements of an Effective Emergency Plan

An effective **emergency response plan** should contain several key elements to ensure a coordinated and efficient response in the event of an emergency:

1. **Hazard Identification**
   The plan must identify potential emergency scenarios that could occur in the workplace (e.g., fire, chemical spill, earthquake).
2. **Roles and Responsibilities**
   Clearly defined roles for employees, management, and emergency responders. Each person should know their specific duties during an emergency.
3. **Evacuation Procedures**
   Detailed instructions on how to evacuate the building safely, including designated exits and assembly points.
4. **Emergency Contact Information**
   A list of emergency contacts, including fire services, medical personnel, and key internal personnel, must be included.
5. **First Aid and Medical Assistance**
   Protocols for providing immediate medical care or first aid in the event of an injury. This includes the location of first aid kits and designated first aid responders.
6. **Communication Protocols**
   Clear communication channels for informing employees about the emergency, including alarms, loudspeakers, or other notification systems.
7. **Training and Drills**
   Regular training and drills to ensure all employees are familiar with the emergency response procedures and know their roles in an emergency.

Ensuring Worker Safety During Emergencies

Ensuring worker safety during emergencies is critical for reducing injuries and fatalities. This can be achieved by:

1. **Providing Personal Protective Equipment (PPE)**
   Ensuring that employees have access to appropriate PPE, such as fire-resistant clothing, respiratory protection, or safety goggles, depending on the type of emergency.
2. **Establishing Safe Evacuation Routes**
   Clearly marking evacuation routes, ensuring they are free of obstacles, and providing adequate signage to guide workers during an emergency evacuation.
3. **Maintaining Emergency Equipment**
   Regularly inspecting and maintaining emergency equipment such as fire extinguishers, first aid kits, and emergency lighting.
4. **Designating Safe Assembly Areas**
   Identifying safe assembly points outside the workplace where employees can gather during an evacuation.

Training and Drills for Emergency Scenarios

Training and drills are essential components of emergency preparedness. Regularly conducted drills help ensure that workers are familiar with emergency procedures, and that they know how to respond quickly and effectively during an emergency. Key components of training and drills include:

- **Regularly Scheduled Drills**: Conducting fire drills, evacuation drills, and other scenario-based exercises at regular intervals.
- **Scenario-Based Training**: Providing workers with hands-on experience in dealing with various emergency scenarios, such as chemical spills or first-aid situations.
- **Post-Drill Evaluation**: After each drill, a review is conducted to evaluate the effectiveness of the response and identify areas for improvement.
- **Refresher Courses**: Providing periodic refresher training to ensure that workers retain critical emergency response knowledge.

Documentation and Communication

Effective documentation and communication are crucial to ensuring that emergency preparedness procedures are clear and accessible. Key aspects include:

- **Written Procedures**: Documenting all emergency response procedures in a clear, easy-to-understand format.
- **Accessibility**: Ensuring that emergency procedures are easily accessible to all employees, including posting them in common areas or on a shared platform.
- **Communication Systems**: Having a communication system in place to notify employees quickly about emergencies (e.g., alarms, emails, SMS alerts).
- **Regular Updates**: Keeping all documentation up to date with any changes to emergency procedures or contact information.

Keeping Emergency Procedures Accessible

Emergency procedures must be easily accessible to all employees at all times. This can be accomplished by:

- **Posting Procedures in Common Areas**: Displaying emergency procedures in high-traffic areas, such as break rooms, hallways, and near exits.
- **Digital Access**: Providing digital copies of emergency procedures on the company's internal network or intranet so employees can easily reference them.
- **Accessible for All Workers**: Ensuring that emergency procedures are available in multiple languages or formats (e.g., large print, audio) to accommodate all employees, including those with disabilities.

Communicating Roles and Responsibilities Clearly

During an emergency, it's essential that all employees understand their roles and responsibilities to ensure a coordinated response. This can be achieved by:

- **Training Programs**: Regular training sessions where roles and responsibilities are clearly communicated to all employees.

- **Emergency Role Assignments**: Assigning specific tasks (e.g., evacuation leader, first-aid responder) to employees before an emergency occurs, so they are ready when needed.
- **Clear Instructions**: Providing clear, concise instructions on what is expected from each individual during an emergency.

Table: Key Elements of Emergency Preparedness

Element	Description	Action/Outcome
Hazard Identification	Identifying potential emergency scenarios (fire, chemical spill, etc.).	Helps prioritize emergency planning and ensure appropriate responses to specific risks.
Roles and Responsibilities	Defining specific roles for employees during an emergency (e.g., evacuation leaders, first-aiders).	Ensures that all employees know their tasks and can respond effectively in an emergency.
Evacuation Procedures	Clear instructions on how to evacuate safely, including routes and assembly points.	Reduces confusion and ensures that employees can evacuate quickly and safely.
Emergency Contact Information	A list of emergency services and key personnel to contact during an emergency.	Facilitates fast communication with external emergency responders and internal key personnel.
First Aid and Medical Assistance	Protocols for providing first aid and medical care.	Ensures quick medical response to injured workers, reducing the impact of incidents.
Training and Drills	Regularly scheduled emergency drills and refresher training for employees.	Prepares employees for emergencies, ensuring they can act quickly and correctly when needed.
Communication Protocols	Communication channels for alerting employees about emergencies.	Ensures that all employees are quickly informed during an

		emergency and can follow proper procedures.

Conclusion

Operational controls and emergency preparedness are crucial aspects of ISO 45001 that help ensure the safety and well-being of employees. By implementing effective **operational controls** and developing robust **emergency preparedness plans**, organizations can manage risks proactively, respond effectively to emergencies, and minimize harm. Regular training, documentation, and clear communication further enhance preparedness, ensuring that all workers know their roles and responsibilities and are equipped to act swiftly in the event of an emergency.

Practical Tools for ISO 45001 Implementation

In order to help implement operational controls and emergency preparedness effectively, the following **Practical Tools** are essential. These templates will assist in creating structured and efficient processes that ensure compliance with ISO 45001 standards, as well as improve workplace safety through organized procedures and response strategies.

---

1. Operational Control Procedures Template

This **Operational Control Procedures Template** helps organizations document and standardize safety measures for specific tasks, processes, or equipment. By implementing operational controls, organizations can reduce hazards, improve compliance, and ensure a safer workplace.

[Company Name] Operational Control Procedure

1. Title of Procedure
[Insert the name of the process or task the procedure will cover (e.g., Machine Guarding, Chemical Handling, etc.)]

2. Objective
To describe the control measures to be followed for ensuring safety during [Insert task/process] in compliance with ISO 45001 standards.

3. Scope
This procedure applies to all employees, contractors, and visitors involved in [Insert task/process], within the department of [Department Name], across all facilities of [Company Name].

4. Responsibilities

- **[Role/Position Name]**: Ensure the implementation and maintenance of this procedure.
- **[Role/Position Name]**: Monitor and enforce compliance with safety measures during [Insert task/process].
- **Employees**: Follow the safety protocols outlined in this procedure and report hazards.

5. Operational Controls

- **Identification of Risks**: [List and describe potential risks associated with the task or process, such as mechanical failure, chemical exposure, etc.]
- **Control Measures**: [Describe the control measures to mitigate each identified risk. For example, machine guards, ventilation, PPE.]
- **Monitoring**: [Explain how the effectiveness of controls will be monitored, such as regular inspections, safety audits, etc.]

6. Procedure Steps

- **Step 1**: [First step in the process, describing the task with safety instructions]
- **Step 2**: [Next step, detailing the required safety precautions]
- **Step 3**: [Continue steps as necessary to complete the procedure, ensuring safety measures are clear at each stage.]

7. Emergency Procedures

- **In Case of [Emergency Type]**: [List steps to follow in the event of an emergency associated with the procedure.]
- **PPE Requirements**: [Specify any personal protective equipment necessary for this task.]

8. Training Requirements

- All employees must be trained on this procedure before performing the task.
- **Training Frequency**: [Insert how often employees should undergo refresher training.]

9. Review and Revision

- **Review Date**: [Date to review the procedure for updates or improvements]
- **Revision History**: [List of changes made to the procedure document and the dates of updates.]

10. Documentation

- [Insert any forms or logs associated with the procedure, such as inspection logs or risk assessments.]

Signed by: [Name, Position]
Date: [Insert Date]

---

2. Emergency Response Plan Template

An **Emergency Response Plan (ERP)** outlines how an organization will respond to various emergency scenarios. This template helps ensure that all employees are prepared for emergencies and understand their roles and responsibilities.

[Company Name] Emergency Response Plan

1. Purpose
To provide a structured, organized approach to handling emergencies, including fire, chemical spills, and medical emergencies, in accordance with ISO 45001 requirements.

2. Scope
This plan applies to all employees, contractors, and visitors on-site at **[Company Name]** facilities. It outlines procedures for responding to emergencies and minimizing harm to people and property.

3. Types of Emergencies

- **Fire Emergency**
- **Chemical Spill**
- **Medical Emergency**
- **Natural Disasters (e.g., Earthquake, Flood)**
- **Workplace Violence**

4. Emergency Response Team (ERT)

- **Team Leader**: [Name & Position]
- **Fire Safety Officer**: [Name & Position]
- **First-Aid Team**: [Names & Positions]
- **Evacuation Leader**: [Name & Position]
- **Communication Officer**: [Name & Position]

5. Emergency Response Procedures

- **Fire Emergency**:
  - **Step 1**: Activate the fire alarm.
  - **Step 2**: Evacuate the building using the nearest fire exits.
  - **Step 3**: Assemble at the designated assembly point.
  - **Step 4**: Contact emergency services (Fire Department) at [Phone Number].
- **Chemical Spill**:
  - **Step 1**: Evacuate the affected area immediately.
  - **Step 2**: Seal off the area to prevent the spread of chemicals.

- **Step 3**: Notify the emergency team and report the chemical spill to authorities.
- **Step 4**: Refer to the Material Safety Data Sheet (MSDS) for the specific chemical to manage the spill safely.

6. Communication Protocols

- **Internal Communication**: Use loudspeakers and safety alarms to communicate the emergency situation.
- **External Communication**: Contact emergency services immediately by phone, giving clear details of the situation.

7. Evacuation Procedures

- **Evacuation Routes**: Clearly marked evacuation routes for all areas of the facility.
- **Assembly Points**: Designated assembly points for employees to gather once safely evacuated.

8. First-Aid and Medical Assistance

- **First-Aid Kits Location**: [Location of first-aid kits on-site].
- **Medical Response**: In case of serious injury, contact emergency medical services immediately. Provide first-aid as needed.

9. Review and Updates

- This emergency response plan will be reviewed annually and updated as necessary.
- Emergency drills will be conducted every [Insert Frequency].

Signed by: [Name, Position]
Date: [Insert Date]

---

3. Incident Reporting and Investigation Form

The **Incident Reporting and Investigation Form** is used to document workplace incidents, accidents, near misses, and unsafe conditions. This form ensures that all incidents are reported, investigated, and corrective actions are taken to prevent future occurrences.

[Company Name] Incident Reporting and Investigation Form

1. Incident Information

- **Incident Type**: [Select: Accident / Near Miss / Unsafe Condition]
- **Date & Time of Incident**: [Insert Date and Time]
- **Location of Incident**: [Insert Location]
- **Reported by**: [Name, Job Title]
- **Involved Persons**: [List names and roles of affected individuals]

2. Incident Description

[Provide a detailed description of the incident, including what happened, how it occurred, and any contributing factors.]

3. Immediate Action Taken

[Describe any immediate actions taken to mitigate the incident or prevent further harm, such as first-aid administered, evacuation, etc.]

4. Incident Classification

- **Severity**: [Minor / Major / Critical]
- **Cause**: [Human Error / Mechanical Failure / Environmental Factors / Other]

5. Investigation

- **Investigation Lead**: [Name, Position]
- **Findings**: [Summarize the investigation findings, identifying the root causes of the incident.]

6. Corrective and Preventive Actions (CAPA)

- **Corrective Action(s)**: [Describe the corrective actions to address the immediate cause of the incident.]
- **Preventive Action(s)**: [Describe the preventive actions to ensure similar incidents do not occur in the future.]
- **Action Deadline**: [Insert Date]

7. Follow-up and Verification

- **Follow-up Actions**: [Outline the steps for verifying that corrective actions were implemented and effective.]
- **Follow-up Date**: [Insert Date]

8. Additional Comments
[Provide any additional relevant details or observations.]

Signed by: [Name, Position]
Date: [Insert Date]

Conclusion

These practical templates—**Operational Control Procedures, Emergency Response Plan**, and **Incident Reporting and Investigation Form**—are essential tools for aligning your organization with ISO 45001 standards. By using these templates, your organization can ensure that safety protocols are effectively documented, employees know how to respond during emergencies, and incidents are thoroughly investigated to improve workplace safety continuously.

Chapter 4

Monitoring, KPIs, and Internal Audits

Introduction

In ISO 45001, **monitoring**, **Key Performance Indicators (KPIs)**, and **internal audits** play pivotal roles in ensuring that the **Occupational Health and Safety Management System (OHSMS)** is not only compliant with the required standards but also continuously improving. These elements help organizations track the effectiveness of their safety management systems, identify areas for improvement, and take corrective actions as necessary.

This chapter explores how to monitor and measure safety performance, the critical role of KPIs in evaluating the system's success, and the importance of conducting internal audits to drive continuous improvement in health and safety practices.

Monitoring and Measurement in ISO 45001

Monitoring and measurement are fundamental activities within the ISO 45001 framework. These processes allow organizations to track their OHSMS performance, ensuring that health and safety objectives are being met and risks are being effectively controlled.

1. **Continuous Monitoring of Safety Activities**
   Regular monitoring of safety-related activities and processes ensures that any issues are identified early. Monitoring activities can include:
    - **Workplace Inspections**: Regular inspections of the work environment to identify potential hazards.
    - **Safety Observations**: Observing workers as they perform tasks to ensure they follow safety protocols.
    - **Incident Tracking**: Monitoring incidents and near misses to identify trends and areas for improvement.

2. **Measuring Safety Performance**
   Measurement involves collecting data related to specific safety outcomes and comparing them to set safety objectives or targets. Some common methods of measurement include:
   - **Accident and Incident Rates**: Tracking the number and types of workplace accidents, injuries, or illnesses.
   - **Safety Training Completion Rates**: Measuring how many employees have completed required safety training.
   - **Compliance Rates**: Monitoring compliance with safety policies, procedures, and legal regulations.
3. **Use of Data for Decision Making**
   The data collected from monitoring and measurement activities help organizations make informed decisions. Regular data analysis can highlight patterns and potential risks, guiding corrective actions and the allocation of resources to high-priority areas.

The Role of KPIs in Evaluating Safety Performance

Key Performance Indicators (KPIs) are quantifiable metrics used to evaluate the effectiveness of an organization's safety management system and the achievement of safety goals. KPIs provide organizations with a way to objectively assess their safety performance and drive improvement. The role of KPIs in ISO 45001 is critical as they provide a clear indication of how well safety objectives are being met.

1. **Setting Measurable Safety Objectives**
   KPIs help define clear, measurable safety objectives aligned with the organization's overall goals. For example:
   - **Reduce the number of workplace injuries by 10% over the next year.**
   - **Ensure 100% of employees complete their safety training by the end of the quarter.**
2. **Common Safety KPIs**
   Some commonly used KPIs in ISO 45001 to measure safety performance include:
   - **Total Recordable Injury Rate (TRIR)**: A measure of the number of injuries or illnesses per 100 employees.

    - **Lost Time Injury Frequency Rate (LTIFR)**: The number of injuries resulting in lost workdays per million hours worked.
    - **Near Miss Frequency Rate**: The number of near misses reported per unit of time or number of employees.
    - **Safety Training Completion Rate**: The percentage of employees who have completed their required safety training.
3. **Tracking Safety Trends and Performance**
   KPIs provide valuable insights into safety trends. By consistently tracking these metrics, organizations can identify emerging risks, monitor the effectiveness of existing controls, and prioritize areas for improvement. KPIs should be reviewed regularly to ensure that safety goals are being met and that the organization's health and safety management efforts are yielding the desired results.
4. **KPIs and Continuous Improvement**
   KPIs are directly tied to the concept of **continuous improvement** in ISO 45001. Organizations use the data gathered from KPIs to identify weak points in their safety management system and take corrective actions. If KPIs show that injury rates are higher than expected or training completion is lower than planned, management can adjust policies, allocate resources, or implement new controls to improve performance.

Importance of Internal Audits for Continuous Improvement

Internal audits are an essential tool for evaluating the effectiveness of the OHSMS and ensuring that it is functioning as intended. These audits are a key part of the continuous improvement cycle in ISO 45001.

1. **What is an Internal Audit?**
   An **internal audit** is a systematic, independent examination of the OHSMS to determine whether it complies with ISO 45001 standards, the organization's own policies, and regulatory requirements. The audit assesses the performance of the system and identifies any areas of non-compliance or opportunities for improvement.
2. **Key Purposes of Internal Audits**
    - **Compliance Checking**: Audits ensure that the organization is in compliance with ISO 45001 and any applicable legal requirements related to occupational health and safety.

- **Identifying Weaknesses**: Audits identify areas where the OHSMS is not performing effectively or where improvements can be made.
- **Corrective Action**: The audit process identifies areas for corrective action, helping the organization make necessary adjustments to address deficiencies.
- **Validation of Effectiveness**: Internal audits validate whether the controls and measures in place are working as intended to reduce risks and improve safety outcomes.
- **Preparing for External Audits**: Audits help organizations prepare for certification or recertification audits by ensuring that they meet all the necessary requirements of ISO 45001.

3. **Internal Audit Process**
   The internal audit process typically follows these steps:
   - **Planning**: Define the scope of the audit and identify which parts of the OHSMS will be audited.
   - **Conducting the Audit**: The auditor collects data, conducts interviews, and reviews documentation to assess compliance.
   - **Reporting**: The findings of the audit are documented, and any non-conformities or areas for improvement are highlighted.
   - **Follow-up**: After the audit, corrective actions are taken to address any identified issues, and the effectiveness of these actions is monitored.
4. **Internal Audits and Continuous Improvement**
   Internal audits are a key mechanism for **continuous improvement** in ISO 45001. By regularly auditing the system, organizations can identify areas where they are falling short of safety goals, ensure compliance, and refine processes. This leads to:
   - **Improved Safety Culture**: Continuous feedback through audits helps reinforce a culture of safety, where employees and management are always looking for ways to improve.
   - **Ongoing Risk Reduction**: Audits help identify new and emerging risks, allowing the organization to adapt its controls and procedures to reduce those risks over time.
   - **Evidence for Decision Making**: The audit findings provide concrete evidence that can inform management decisions about safety policies, procedures, and resource allocation.

Conclusion

Monitoring, KPIs, and internal audits are critical components of ISO 45001 that help organizations continuously evaluate and improve their Occupational Health and Safety Management System. **Monitoring** ensures that safety practices are being implemented effectively; **KPIs** provide a measurable way to track safety performance; and **internal audits** offer an independent review of the system's effectiveness, ensuring that it remains compliant and continuously improves.

Together, these elements create a robust framework for maintaining a safe workplace, meeting ISO 45001 requirements, and fostering a culture of **continuous improvement** in health and safety. By regularly evaluating and refining safety practices, organizations can prevent incidents, reduce risks, and provide a safer working environment for all employees.

Key Concepts

Defining Key Performance Indicators (KPIs)

Key Performance Indicators (KPIs) are measurable values used to evaluate the success of an organization in achieving specific safety-related objectives within the framework of ISO 45001. KPIs provide a clear, quantifiable way to monitor the performance of an **Occupational Health and Safety Management System (OHSMS)** and ensure that safety goals are being met.

KPIs in ISO 45001 should:

- Be directly linked to **safety objectives** and help evaluate the effectiveness of the OHSMS.
- Provide a **baseline for comparison** over time, allowing organizations to track progress and identify trends.
- Be used to inform **decision-making**, providing management with the data needed to adjust safety protocols, allocate resources, and address safety issues.

For example, a KPI could track the **number of workplace injuries** over a specific period, and compare that to previous periods or organizational goals.

Common KPIs in Occupational Health and Safety

Some commonly used KPIs in the **Occupational Health and Safety (OHS)** field, specifically aligned with ISO 45001, include:

1. **Total Recordable Injury Rate (TRIR)**
   TRIR measures the number of recordable injuries per 100 full-time employees. This includes all work-related injuries or illnesses that require medical treatment or result in lost workdays.
2. **Lost Time Injury Frequency Rate (LTIFR)**
   LTIFR measures the frequency of injuries resulting in lost workdays, expressed per one million hours worked. It provides insight into the severity of workplace incidents.
3. **Near Miss Frequency Rate**
   This KPI tracks the number of near misses (incidents where no injury occurred but could have led to one) reported in a given period. A higher near miss rate indicates that the organization is effectively identifying potential hazards.
4. **Safety Training Completion Rate**
   This tracks the percentage of employees who have completed mandatory safety training. It is critical for ensuring that employees are equipped with the knowledge to recognize and mitigate risks.
5. **Accident Severity Rate**
   This KPI measures the severity of accidents by calculating the total number of lost days due to injuries divided by the total number of injuries. It reflects how severe injuries are in relation to their frequency.
6. **Compliance Rate**
   This KPI measures the extent to which the organization complies with health and safety laws and regulations. It indicates how well the organization adheres to legal obligations and internal safety standards.
7. **Employee Safety Engagement**
   This measures the level of employee involvement in safety programs, which can include participation in safety meetings, hazard reporting, and involvement in safety committees.

By using these KPIs, organizations can monitor how well their safety management system is performing and identify areas that need attention or improvement.

Setting Realistic and Measurable Targets

When setting KPIs, it is essential that the targets are both **realistic** and **measurable**. To ensure effectiveness, targets should be:

- **Specific**: Clearly define what is being measured (e.g., reduce TRIR by 5%).
- **Measurable**: Ensure that the target can be tracked and quantified (e.g., number of lost-time injuries).
- **Achievable**: Set a goal that is challenging but attainable with available resources and within the time frame.
- **Relevant**: Align the targets with the organization's safety objectives and overall business goals.
- **Time-bound**: Set a deadline or a timeframe within which the target should be achieved.

By ensuring targets are realistic and measurable, organizations increase the likelihood of success and can track progress over time.

Monitoring and Measuring Safety Performance

Monitoring and measuring safety performance are key components of **ISO 45001** that ensure the OHSMS is functioning effectively and helping to reduce workplace risks. This process allows the organization to track progress against set safety objectives, such as reducing injury rates or increasing safety training participation.

The key steps involved in monitoring and measuring safety performance include:

- **Data Collection**: Gather data from a variety of sources, including accident reports, inspections, and employee feedback.
- **Regular Review**: Continuously review safety metrics to ensure the OHSMS is performing as expected.
- **Tracking Progress**: Use KPIs to track improvements over time. This allows organizations to identify trends and make necessary adjustments.

- **Analysis and Reporting**: Analyze data to understand the causes of safety incidents and determine areas for improvement. Reporting should be clear, concise, and focused on actionable insights.

Tools for Tracking Safety Metrics

To effectively track safety metrics and KPIs, organizations need the right tools. Some of the tools that can be used to track safety performance include:

1. **KPI Tracking Spreadsheet**
   A **KPI Tracking Spreadsheet** is a tool that allows organizations to log and track various safety metrics over time. This spreadsheet includes:
   - Columns for the specific safety metric (e.g., TRIR, LTIFR, etc.)
   - Time periods (e.g., monthly, quarterly, annually)
   - Actual performance data and comparisons with targets or previous periods
   - Notes for any incidents, corrective actions, or comments.
2. **Safety Management Software**
   Specialized software can be used to track safety performance and analyze trends. These platforms provide real-time data, making it easier to identify risks and take action quickly.
3. **Dashboards and Visualization Tools**
   Dashboards and data visualization tools allow organizations to display KPIs in a clear, visual format. These tools often include charts and graphs, making it easier for managers and employees to track performance and make data-driven decisions.

Conducting Safety Inspections and Risk Assessments

Safety inspections and risk assessments are essential for identifying and mitigating workplace hazards. Regular inspections ensure that safety controls are in place and functioning correctly, while risk assessments evaluate the likelihood and impact of potential hazards.

1. **Safety Inspections**
   Regular inspections help identify hazards in the workplace. Inspections can focus on:
   - Equipment safety

   - Workplace environment (e.g., lighting, ventilation)
   - PPE usage and maintenance
   - Fire safety measures
   - Chemical storage and handling practices
2. **Risk Assessments**
   Risk assessments are systematic processes used to identify hazards and assess their potential risks. The assessment should consider:
   - The **likelihood** of the hazard occurring
   - The **severity** of the potential outcome if the hazard occurs
   - The **existing control measures** in place
   - The need for additional controls or safety measures.

Internal Audits and Non-Conformance

Internal audits are an essential tool for ensuring that the **OHSMS** is functioning effectively and that the organization remains compliant with ISO 45001. Audits help identify any non-conformities (areas where the system is not meeting requirements) and provide opportunities for improvement.

1. **The Purpose of Internal Audits**
   The main purposes of conducting internal audits include:
   - **Ensuring Compliance**: Verifying that the organization complies with ISO 45001 standards, as well as legal and regulatory requirements.
   - **Identifying Gaps**: Detecting weaknesses or inefficiencies in the safety management system.
   - **Providing Evidence for Certification**: Internal audits serve as evidence that the system is functioning and continuously improving, which is essential for certification audits.
2. **Steps for Conducting an Effective Internal Audit**
   The audit process generally involves:
   - **Planning**: Defining the scope of the audit, including which processes or areas will be audited.
   - **Collecting Evidence**: Gathering data through document reviews, interviews, and observations of work processes.
   - **Identifying Non-Conformances**: Documenting any areas where the OHSMS is not meeting ISO 45001 requirements or internal policies.

 - **Reporting**: Presenting audit findings to management, including recommendations for corrective actions.
 - **Follow-up**: Ensuring corrective actions are implemented and verifying their effectiveness.

3. **Handling Non-Conformance and Corrective Actions**
   When non-conformities are identified, corrective actions should be taken promptly. The process includes:
   - **Root Cause Analysis**: Determining the underlying cause of the non-conformance to prevent recurrence.
   - **Corrective Action Plan**: Developing an action plan that outlines how the issue will be addressed and corrected.
   - **Implementation**: Carrying out the corrective actions within an agreed timeframe.
   - **Verification**: Following up to ensure that corrective actions have effectively addressed the problem and that the system is now compliant.

Practical Tools

---

1. KPI Tracking Spreadsheet

A **KPI Tracking Spreadsheet** is a simple tool that helps organizations log and track safety-related KPIs and performance metrics. It allows users to compare current performance with set targets and identify trends over time.

[Company Name] KPI Tracking Spreadsheet

KPI	Target	Actual Performance	Date	Notes
Total Recordable Injury Rate (TRIR)	2.5	3.1	Q1 2026	Slight increase due to equipment failure incident.

Lost Time Injury Frequency Rate (LTIFR)	0.5	0.4	Q1 2026	Goal met, no LTIs reported.
Safety Training Completion Rate	100%	85%	Q1 2026	Need to complete safety training for 15 employees.

2. Internal Audit Checklist

An **Internal Audit Checklist** ensures that the audit process is comprehensive and consistent. It includes key points that auditors should review to ensure that the OHSMS is compliant with ISO 45001.

[Company Name] Internal Audit Checklist

Audit Area	Audit Criteria	Status (Compliant/Non-Compliant)	Comments
Leadership and Worker Participation	Evaluate management commitment to safety and employee involvement.	Compliant	Management signed OHS policy.
Risk Assessment and Control Measures	Check if risks are identified, assessed, and controlled.	Non-Compliant	Risk assessments not updated in 6 months.
Emergency Preparedness	Review emergency response plans and drills.	Compliant	Emergency drill conducted last month.

3. Corrective Action Request (CAR) Template

A **Corrective Action Request (CAR) Template** is used to document non-conformances identified during audits, inspections, or incident investigations and track the actions taken to correct them.

[Company Name] Corrective Action Request (CAR) Template

CAR Number	[CAR-001]	Date Issued	[Insert Date]
Non-Conformance	Risk assessments are not updated regularly.	**Responsible Person**	[Name & Position]
Root Cause	Lack of regular review of risk assessments.	**Corrective Action**	Update risk assessments every 6 months.
Target Date	[Insert Date]	**Follow-up Date**	[Insert Date]
Verification	[Audit/Inspection Date]	**Status**	Pending/Resolved

Conclusion

By using these **Practical Tools**—the **KPI Tracking Spreadsheet**, **Internal Audit Checklist**, and **Corrective Action Request Template**—organizations can effectively monitor and measure safety performance, identify areas for improvement, and ensure continuous enhancement of the **Occupational Health and Safety Management System (OHSMS)**. These tools help track progress against KPIs, ensure compliance with ISO 45001, and implement corrective actions that lead to a safer and more efficient workplace.

Chapter 5

Certification Audit Preparation and Success Strategy

Introduction

Achieving **ISO 45001 certification** is the final milestone in the journey of establishing an effective **Occupational Health and Safety Management System (OHSMS)**. However, before certification can be awarded, the organization must undergo a **certification audit**. This chapter provides guidance on preparing for the certification audit and emphasizes the importance of being thoroughly prepared to ensure a smooth and successful audit process. Proper preparation is crucial not only to meet the requirements of the standard but also to demonstrate the organization's commitment to ongoing safety and improvement.

The Final Step: Preparing for the Certification Audit

The **certification audit** is the external evaluation of an organization's OHSMS, conducted by an accredited third-party certification body. The purpose of this audit is to assess whether the organization's safety management system meets the requirements of ISO 45001 and is effective in managing health and safety risks.

Steps for Preparing for the Certification Audit:

1. **Review ISO 45001 Requirements**
   Before the audit, review the **ISO 45001 standard** thoroughly to ensure all requirements are met. This includes understanding the specific clauses of the standard that relate to leadership, risk management, operational controls, emergency preparedness, and continual improvement.
2. **Ensure Documentation Is Complete**
   All policies, procedures, and records must be in place and up to date. The auditor will review these documents to verify that the OHSMS aligns with the ISO 45001 framework. Make sure:

- **Safety policies** are clear, comprehensive, and signed by top management.
- **SOPs** are documented and available for all key safety processes.
- **Risk assessments** and **corrective actions** are recorded and up to date.
- **Employee training** records are complete and show participation in relevant safety programs.

3. **Conduct an Internal Audit**
   Conduct a **final internal audit** of your OHSMS to ensure that all aspects of the system are functioning properly and are aligned with ISO 45001 requirements. The internal audit should identify any potential weaknesses or areas of non-compliance that need to be addressed before the external audit.
4. **Engage and Train Employees**
   Make sure all employees are familiar with the OHSMS and know what to expect during the audit. Key personnel should be trained on:
   - **Audit processes** and what auditors may ask.
   - **Safety policies** and procedures.

- **Their roles and responsibilities** within the safety management system.

5. **Corrective Actions**
If the internal audit has identified non-conformities, take corrective actions promptly. Correct any gaps or deficiencies in processes, procedures, or documentation. Ensure these corrective actions are well-documented and that their implementation is complete before the certification audit.
6. **Pre-Audit Review**
Conduct a **pre-audit review** where you can walk through the OHSMS with your team, conducting a mock audit to ensure the system is fully functional and in compliance with ISO 45001. This mock audit will help to identify any last-minute issues that could arise during the official certification audit.

The Importance of a Well-Prepared Audit

Being well-prepared for the **certification audit** is critical to ensure that the audit is smooth, efficient, and successful. Proper preparation helps the organization demonstrate its commitment to workplace safety and allows the auditors to see that the OHSMS is functioning effectively.

1. **Confidence in System Integrity**
A well-prepared OHSMS indicates to auditors that the organization is committed to maintaining a high standard of safety. If everything is in order—documents, procedures, employee training, and operational controls—this will make the audit process smoother and demonstrate to auditors that the organization is serious about safety. Preparation also gives the organization confidence that they are meeting the requirements of ISO 45001.
2. **Timely Identification and Resolution of Issues**
A thorough preparation process allows the organization to identify and resolve any issues or non-conformities ahead of the audit. By addressing gaps before the auditor arrives, the organization ensures that there are no delays or setbacks during the audit process.
3. **Efficiency During the Audit**
When the organization is well-prepared, the audit process is much more

efficient. Auditors will appreciate being provided with well-organized documentation, clear records, and well-trained personnel ready to answer questions. This reduces the audit time and avoids unnecessary delays or confusion.

4. **Positive Impression with Auditors**
   A well-prepared organization leaves a positive impression on auditors, showing that they take safety management seriously and that they are committed to the continuous improvement of their OHSMS. A positive audit experience may even lead to valuable feedback that can further enhance the safety system.
5. **Increased Likelihood of Certification Success**
   Being prepared increases the likelihood of a successful certification audit. Auditors will be able to quickly verify compliance with ISO 45001, and the organization will demonstrate that its OHSMS is fully functional. If the organization is adequately prepared, auditors are more likely to recommend certification with minimal issues.
6. **Identifying Opportunities for Improvement**
   A well-prepared audit can also uncover areas where the organization can further enhance its OHSMS. ISO 45001 encourages continual improvement, and the audit process can provide valuable insights into how safety practices can be refined, even if the system is already compliant.
7. **Building a Safety-First Culture**
   Preparing for the certification audit also promotes a **safety-first culture** within the organization. It encourages employees to take safety seriously, knowing that the organization is going through a certification process. It also reinforces the importance of continuous improvement in safety management, which is at the core of ISO 45001.

Conclusion

The certification audit is a crucial step in obtaining ISO 45001 certification and demonstrating an organization's commitment to **occupational health and safety**. Preparing for the audit requires a comprehensive approach, including reviewing documentation, conducting internal audits, engaging employees, and ensuring corrective actions are taken.

A well-prepared audit process will lead to a smoother and more efficient audit, reducing the likelihood of non-conformities and improving the chances of a successful certification. In addition, preparation ensures that the organization is ready for continuous improvement, an essential aspect of maintaining ISO 45001 certification and fostering a safe and healthy work environment.

Key Concepts

Preparing for the Certification Audit

The **certification audit** is the final step in the process of achieving ISO 45001 certification. This audit is conducted by an accredited third-party auditor who will evaluate whether the organization's **Occupational Health and Safety Management System (OHSMS)** meets the requirements of ISO 45001 and whether it is functioning effectively. Preparing for this audit is crucial for ensuring that the organization is fully ready to demonstrate compliance and that the process is smooth and efficient.

Preparation includes:

- **Reviewing ISO 45001 Requirements**: Ensure that the organization is fully aligned with the standard.
- **Internal Audits**: Conduct a final internal audit to ensure compliance and identify any last-minute issues.
- **Employee Training**: Ensure that employees understand the system and know their roles during the audit.
- **Documentation Review**: Check that all safety policies, procedures, risk assessments, and training records are up to date and readily available.

A well-prepared OHSMS will make the audit process more efficient and increase the likelihood of passing the certification with minimal corrective actions required.

Understanding the Certification Process

The **certification process** involves several stages, with each stage focusing on evaluating the readiness of the organization's OHSMS. Typically, the process involves the following steps:

1. **Application**: The organization applies for certification by selecting an accredited certifying body.
2. **Stage 1 Audit (Documentation Review)**: In this stage, the auditor reviews the organization's documentation (e.g., policies, procedures, risk assessments, etc.) to ensure it aligns with ISO 45001 requirements.
3. **Stage 2 Audit (Full Audit)**: This stage involves a comprehensive on-site audit where the auditor evaluates the implementation of the OHSMS. They will review records, conduct interviews with employees, and inspect the workplace to assess compliance and effectiveness.
4. **Certification Decision**: After completing the audits, the certifying body will issue a certification decision. If successful, the organization will receive ISO 45001 certification. If there are non-conformities, the organization will need to address them before certification is granted.

Key Areas the Auditor Will Review

During the **Stage 2 audit**, auditors will focus on several key areas to assess whether the organization meets the requirements of ISO 45001. These areas include:

1. **Leadership and Worker Participation**
   The auditor will assess whether top management is actively involved in the OHSMS and whether employees are engaged in safety processes and decision-making.
2. **Risk Management and Hazard Identification**
   The auditor will review how the organization identifies hazards, assesses risks, and implements controls to prevent incidents.
3. **Compliance with Legal Requirements**
   The auditor will ensure that the organization complies with local, national, and international safety laws and regulations.
4. **Documentation and Record-Keeping**
   The auditor will check if the organization has well-documented policies,

procedures, and records, including training records, risk assessments, and audit results.

5. **Emergency Preparedness and Response**
   The auditor will evaluate the organization's emergency response plans and their effectiveness in managing emergency situations.
6. **Internal Audits and Reviews**
   The auditor will verify that internal audits and management reviews are conducted regularly and that corrective actions are implemented to address any identified weaknesses.

Ensuring Compliance with All ISO 45001 Clauses

ISO 45001 consists of several clauses that outline the requirements for a comprehensive OHSMS. Ensuring compliance with all these clauses is essential for passing the certification audit. Key clauses that auditors will focus on include:

1. **Clause 4 (Context of the Organization)**: Understanding internal and external factors that affect the OHSMS, including legal, social, and organizational context.
2. **Clause 5 (Leadership and Worker Participation)**: Ensuring management's commitment to safety and fostering worker participation in decision-making processes.
3. **Clause 6 (Planning)**: Developing processes for risk assessment, setting objectives, and determining the necessary actions to achieve those objectives.
4. **Clause 7 (Support)**: Ensuring the availability of resources, training, and communication to support the OHSMS.
5. **Clause 8 (Operation)**: Implementing operational controls to manage risks and ensuring compliance with safety procedures.
6. **Clause 9 (Performance Evaluation)**: Monitoring the performance of the OHSMS through audits, reviews, and corrective actions.
7. **Clause 10 (Improvement)**: Ensuring continuous improvement of the OHSMS based on feedback, audits, and performance evaluations.

Creating an Audit Readiness Checklist

An **audit readiness checklist** is a tool that helps ensure the organization is fully prepared for the certification audit. It includes all the necessary tasks and documentation that need to be reviewed, updated, and organized before the audit.

Audit Readiness Checklist should include:

- **Documentation Review**: Ensure that all policies, procedures, records, and training documents are complete and accessible.
- **Internal Audit Results**: Ensure that all non-conformities from the internal audit have been addressed.
- **Employee Preparedness**: Ensure that employees are trained and understand the audit process.
- **Physical Workplace Inspections**: Check that the workplace is safe and meets the standards outlined in the OHSMS.
- **Emergency Procedures**: Ensure that emergency preparedness plans are updated and accessible.

Internal Audit Findings vs. Certification Audit

The **internal audit** is an important tool for preparing for the certification audit. Internal audits are conducted before the certification audit to identify areas for improvement and ensure compliance with ISO 45001. Any findings from the internal audit should be addressed before the external certification audit.

Key Differences Between Internal and Certification Audits:

- **Scope**: Internal audits focus on evaluating the effectiveness of the OHSMS within the organization, while the certification audit is a formal evaluation by a third-party auditor to determine if the organization meets the ISO 45001 requirements.
- **Objective**: Internal audits are designed to identify non-conformities and areas for improvement, while certification audits assess whether the organization complies with the ISO 45001 standard.

Preparing Documents and Evidence for Review

During the certification audit, auditors will review all documents and evidence related to the OHSMS. It is essential to prepare and organize the following documents for review:

- **Safety Policy and Objectives**: The organization's safety goals and management commitment.
- **Risk Assessments**: Documentation of risk identification, assessments, and control measures.
- **Training Records**: Evidence that employees have completed required safety training and are competent in safety protocols.
- **Internal Audit Reports**: Records of internal audits and corrective actions taken.
- **Management Review Records**: Evidence that top management reviews the OHSMS and makes decisions based on performance evaluations.

Strategies for Audit Success

1. **Be Prepared and Organized**
   Ensure that all documentation is up to date and accessible. A well-organized system makes it easier for auditors to evaluate the OHSMS and helps demonstrate that the organization is committed to safety.
2. **Engage Employees in the Process**
   Employees should understand their roles during the audit and be ready to answer questions about safety practices, procedures, and their involvement in safety programs.
3. **Ensure Management Commitment**
   Top management should actively participate in the audit process. Their involvement demonstrates the organization's commitment to health and safety and sets a positive tone for the audit.
4. **Be Transparent**
   Be honest and transparent with auditors. If there are areas of non-compliance, it's better to acknowledge them and demonstrate a commitment to corrective actions rather than trying to hide issues.

Creating a Positive Audit Experience

Creating a positive audit experience can significantly impact the outcome of the certification process. Here are some tips to create a positive experience for both the auditors and the organization:

- **Maintain Open Communication**: Encourage open dialogue between the auditors and the organization throughout the process. Clear communication helps avoid misunderstandings and ensures the audit is conducted efficiently.
- **Be Professional and Courteous**: Treat auditors with respect, listen carefully to their feedback, and respond professionally to their questions and observations.
- **Demonstrate a Commitment to Safety**: Show auditors that safety is a priority at all levels of the organization, from top management to front-line workers.

Responding to Auditor's Questions Effectively

1. **Be Honest and Direct**
   If auditors ask a question about a non-conformity or a safety issue, provide an honest and clear response. If an issue exists, explain how you are working to address it.
2. **Provide Supporting Evidence**
   Be ready to provide documentation and evidence to back up your responses, such as training records, safety reports, or corrective action plans.
3. **Stay Calm and Professional**
   Auditors may ask probing or difficult questions. Stay calm, be professional, and answer questions to the best of your ability. If you don't know the answer, offer to follow up after the audit.

Managing Last-Minute Compliance Gaps

It's common to discover small compliance gaps just before or during the certification audit. The key to managing these gaps is:

- **Address Minor Issues Quickly**: If small issues are discovered, take immediate corrective actions and document them.
- **Be Transparent**: If a larger issue is identified and cannot be fixed immediately, explain the issue to the auditors and provide a clear action plan with timelines for resolution.
- **Demonstrate Commitment to Continuous Improvement**: Show auditors that you are committed to addressing any gaps and continuously improving your OHSMS, even after certification.

Conclusion

Certification audit preparation is crucial to successfully achieving ISO 45001 certification. By understanding the certification process, reviewing key areas auditors will focus on, preparing necessary documents, and having effective strategies for audit success, organizations can navigate the audit process smoothly. A **well-prepared audit** demonstrates the organization's commitment to workplace safety and health, leading to a positive audit outcome and certification success.

Offer Information

Want to get the full document?

We are offering the Word document version of this book at $20 for those who want to take the content with them wherever they go. For more information and to receive your copy, please contact us at:

Email: standardbookshub@gmail.com

www.ingramcontent.com/pod-product-compliance
Lightning Source LLC
LaVergne TN
LVHW070856210926
848482LV00016B/955